Capacitive Sensors

Capacitive Sensors
Design and Applications

Larry K. Baxter

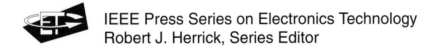

IEEE Press Series on Electronics Technology
Robert J. Herrick, Series Editor

Published under the Sponsorship
of the IEEE Industrial Electronics Society

**IEEE
PRESS**

The Institute of Electrical and Electronics Engineers, Inc.
New York

This book may be purchased at a discount from the publisher when
ordered in bulk quantities. Contact:

IEEE PRESS Marketing
Attn: Special Sales
P.O. Box 1331
445 Hoes Lane
Piscataway, NJ 08855-1331
Fax: (908) 981-9334

For more information about IEEE PRESS products, visit the IEEE Home Page:
http://www.ieee.org/

Printed in the United States of America

10   9   8   7   6   5   4   3   2   1

ISBN 0-7803-1130-2
IEEE Order Number: PC5594

Library of Congress Cataloging-in-Publication Data
Baxter, Larry K.
    Capacitive sensors: design and applications / Larry K. Baxter.
        p.   cm.
    Includes bibliographical references and index.
    ISBN 0-7803-1130-2
    1. Electronic instruments—Design and construction.
    2. Capacitance meters—Design and construction.   I.   Title.
TK7870.B3674   1997
621.37'42—dc20   96-15907
                                                                          CIP

Contents

Preface

Capacitive sensors can solve many different types of sensing and measurement problems. They can be integrated into a printed circuit board or a microchip and offer noncontact sensing with nearly infinite resolution. They are used for rotary and linear position encoding, liquid level sensing, touch sensing, sensitive micrometers, digital carpenter's levels, keyswitches, light switches and proximity detection. Your telephone and tape recorder probably use electret microphones with capacitive sensing, and your car's airbag may be deployed by a silicon accelerometer which uses capacitive sensing. The use of capacitive sensors is increasing rapidly as designers discover their virtues.

Capacitive sensors can be unaffected by temperature, humidity, or mechanical misalignment, and shielding against stray electric fields is simple compared to shielding an inductive sensor against magnetic disturbances. Capacitive transducer accuracy is excellent, as the plate patterns which determine accuracy can be reproduced photographically with micron precision. The technology is easily integrated, and is displacing traditional silicon-based transducers using piezoresistive and piezoelectric effects, as the sensitivity and stability with temperature are ten times better. Capacitive sensors consume very little power; battery life for small portable products may be several years.

Several factors inhibit the use of capacitive sensors, including the specialized circuits needed and the lack of understanding of the technology, including a widely held superstition that capacitive sensors are nonlinear and cannot operate at extremes of humidity. But many products which embed capacitive encoders have been successful in the market. This book surveys different types of sensors and shows how to build rugged, reliable capacitive sensors with high accuracy and low parts cost. Several product designs which use capacitive sensors are analyzed in detail.

The book is organized as follows.

- **Basics** This section covers theoretical background, different electrode geometries, and basic circuit designs.

- **Applications** Four different uses of capacitive sensors are presented: micrometers, proximity detectors, motion encoders, and some miscellaneous sensors. The theory behind these different uses is discussed.
- **Design** The electrode configurations and the basic circuit designs which were briefly discussed in the Basics section are more fully explored here.
- **Products** This section presents design details of several different products which use capacitive sensors.

Thanks are due to John Ames and Rick Grinnell, who helped with writing and review; Katie Gardner, who handled editing, research, and graphics; and my wife Carol, who helped in many other ways.

Larry K. Baxter
Gloucester, Massachusetts

Capacitive Sensors

1

Introduction

Capacitive sensors electronically measure the capacitance between two or more conductors in a dielectric environment, usually air or a liquid. A similar technique is electric field measurement, where the electrostatic voltage field produced by conductors in a dielectric environment is picked up by a probe and a high impedance amplifier.

A high frequency excitation waveform is normally used, as the reactive impedance of small plates in air can be hundreds of megohms at audio frequencies. Excitation frequencies of 100 kHz or more reduce the impedance to a more easily handled kilohm range.

Some systems use the environment as a return path for capacitive currents; a person, for example, is coupled capacitively to the building or the earth with several hundred pF. More often, the return path is a wire.

The range of applications is extreme. At the small end, silicon integrated circuits use the capacitance between micromachined silicon cantilevers to measure micrometer displacements. The dimensions are on the order of 100 μm, and the gap can be a μm or less. At the large end, capacitive personnel and vehicle detectors have plate dimensions of several meters. Capacitive sensors have been used for video-rate signals in one of the early video disk storage technologies. At the sensitive end, one researcher [Jones, p. 589] reports on an instrument with movement sensitivity of 10^{-11} mm.

1.1   SENSOR TRENDS

1.1.1   Market

The world market for sensors [*Control Engineering,* 1993] is over $2.3 billion, growing at an annual rate of 10%. Consumer products, including imaging sensors, make up 64% of the market, with accelerometers accounting for 19% and a variety of specialized sensors completing the total.

Venture Development Corp., Natick, MA, (508)653-9000, in *The U.S. Market for Proximity, Photoelectric, and Linear Displacement Sensors* (1995), forecasts that market segment's size to grow from $429.6 million in 1995 to $548 million in 1999. U.S. consumption in 1999 is forecasted as

Magnetic	1.2%
Ultrasonic	1.9%
Capacitive	3.1%
Magnetic-actuated	4.7%
Inductive	42.2%
Photoelectric	46.9%

1.1.2   Technology

Sensor technology is in the process of a slow migration from discrete "dumb" instruments, expensive and inflexible, to smart, self-calibrating, silicon-based units, and the measurement method of choice for silicon devices as well as for discrete instruments is moving from a variety of transducer technologies, such as magnetic, optical, and piezoelectric, to capacitive.

Capacitive sensors can be used for many different applications. Simple sensors are used for go-no go gaging such as liquid level in reservoirs, where their ability to detect the presence of a dielectric or conductor at a distance allows them to work through a nonconducting window. Similar devices (StudSensor) can find wood studs behind plaster walls. Very high analog accuracy and linearity can be achieved with two- or three-plate systems with careful construction, or multiple-plate geometries can be used with digital circuits to substitute digital precision for analog precision.

Capacitive sensors are also useful for measuring material properties. Materials have different values of dielectric constant as well as dielectric loss, and the values of both properties (particularly dielectric loss, or "loss tangent") change with temperature and frequency to give a material a characteristic signature which can be measured at a distance. Appendix 2 lists the dielectric properties of different materials vs. frequency.

1.1.3   Integration

Tools are available to fabricate many different types of sensors on silicon substrates, including micromachining which etches silicon wafers into sensor structures like diaphragms and cantilevered beams (Chapter 15). The decision to integrate a sensor can be done in stages, with the sensor and its electronics integrated together or separately, and usually is considered when the volume reaches 250,000 per year for total integration. Advantages of integration on silicon [Keister, 1992] are

- Improved sensitivity
- Differential sensors compensate parasitic effects
- Temperature compensation
- Easier A-D conversion
- Batch fabrication for lower cost

- Improved reliability
- Size reduction
- Less assembly required
- Lower parasitic capacitances for smaller capacitive electrodes

Capacitive technology is particularly useful for integration, as it is temperature-stable and problems of parasitic capacitance, lead length, and high impedance nodes are more easily handled on a silicon chip than on a PC board.

1.1.4   Intelligence

Smart sensors increase their value to the system by handling low level tasks such as calibration, self-test, and compensation for environmental factors. The gage factor of smart sensors can be scaled automatically so that one sensor type can be used in many locations, and sensor-resident software may help with communications by managing its end of a sensor data bus.

1.1.5   Sensor bus

Most sensors use point-to-point connections to the host, with contact closure or TTL level digital signaling or 4-20 mA current loop or 1-5 V voltage signaling. As the sensors proliferate in an application, the difficulty of wiring and testing complex systems encourages a single-shared-wire bus connection. One effort to establish smart sensor bus standards is pursued by the IEEE Instrumentation and Measurement Society TC-9 committee [Ajluni, 1994]. This group is drafting "Microfabricated Pressure and Acceleration Technology." A joint IEEE and National Institute of Science and Technology group is developing guidelines for standardization of protocol and interface standards for system designers and manufacturers.

Another proposal for a smart sensor bus is Fieldbus, from the Smar Research Corp. of Brazil [Ajluni, 1994]. Fieldbus works with WorldFip and ISP protocols, and is defined for all layers of the ISO seven-layer communications structure.

1.2   APPLICATIONS OF CAPACITIVE SENSORS

Some different applications listed below show the variety of uses for capacitive sensors.

1.2.1   Proximity sensing

Personnel detection   Safety shutoff when machine operator is too close.

Light switch   A capacitive proximity sensor with 1 m range can be built into a light switch for residential use.

Vehicle detection   Traffic lights use inductive loops for vehicle detection. Capacitive detectors can also do the job, with better response to slow-moving vehicles.

1.2.2   Measurement

Flow   Many types of flow meters convert flow to pressure or displacement, using an orifice for volume flow or Coriolis effect force for mass flow [Jones, 1985]. Capacitive sensors can then measure the displacement directly, or pressure can be converted to a displacement with a diaphragm.

Pressure   With gases or compressible solids a pressure change may be measured directly as a dielectric constant change or a loss tangent change.

Liquid level   Capacitive liquid level detectors sense the liquid level in a reservoir by measuring changes in capacitance between parallel conducting plates which are immersed in the liquid, or applied to the outside of a nonconducting tank.

Spacing   If a metal object is near a capacitor electrode, the coupling between the two is a very sensitive way to measure spacing.

Scanned multiplate sensor   The single-plate spacing measurement concept above can be extended to contour measurement by using many plates, each separately addressed. Both conductive and dielectric surfaces can be measured.

Thickness measurement   Two plates in contact with an insulator will measure the thickness if the dielectric constant is known, or the dielectric constant if the thickness is known.

Ice detector   Airplane wing icing can be detected using insulated metal strips in wing leading edges.

Shaft angle or linear position   Capacitive sensors can measure angle or position with a multiplate scheme giving high accuracy and digital output, or with an analog output with less absolute accuracy but simpler circuitry.

Balances   A spring scale can be designed with very low displacement using a capacitive sensor to measure plate spacing. A similar technique can measure mass, tilt, gravity, or acceleration.

1.2.3   Switches

Lamp dimmer   The standard metal-plate soft-touch lamp dimmer uses 60 Hz excitation and senses the capacitance of the human body to ground.

Keyswitch   Capacitive keyswitches use the shielding effect of the nearby finger or a moving conductive plunger to interrupt the coupling between two small plates.

Limit switch   Limit switches can detect the proximity of a metal machine component as an increase in capacitance, or the proximity of a plastic component by virtue of its increased dielectric constant over air.

1.2.4   Communications

Wireless   Data are capacitively coupled across a short gap in a device which
datacomm   replaces the optical isolator.
                    RF propagation in the near field can be sensed by a receiver with a capacitive plate antenna.

1.2.5   Computer graphic input

***x-y* tablet**   Capacitive graphic input tablets of different sizes can replace the computer mouse as an x-y coordinate input device; finger-touch-sensitive, z-axis-sensitive, and stylus-activated devices are available.

2
Electrostatics

Electrostatics is the study of nonmoving electric charges, electric conductors and dielectrics, and DC potential sources. Most capacitive sensors use simple planar parallel-plate geometry and do not require expertise in electrostatics, but the field is reviewed here for background. A standard text in electrostatics such as Haus and Melcher [1989] can be consulted for more detail. Some applications for capacitive sensors may use nonplanar electrode geometry; these may need more extensive electrostatic field analysis. Analysis tools presented here include closed-form field solutions, electric field sketching, Teledeltos paper simulation and finite element modeling.

2.1   APPROXIMATIONS

Real-world capacitive sensor designs involve moving charges, partially conducting surfaces, and AC potential sources. For an accurate analysis of the fields and currents that make up a capacitive sensor, Maxwell's equations relating electric and magnetic fields, charge density, and current density should be used. But a simplifying approximation which ignores magnetic fields is almost always possible with insignificant loss of accuracy. Systems in which this approximation is reasonable are defined as electroquasistatic [Haus, p.66]:

> Maxwell's equations describe the most intricate electromagnetic wave phenomena. Of course, the analysis of such fields is difficult and not always necessary. Wave phenomena occur on short time scales or at high frequencies that are often of no practical concern. If this is the case the fields may be described by truncated versions of Maxwell's equations applied to relatively long time scales and low frequencies. We will find that a system composed of perfect conductors and free space is electroquasistatic [if] an electromagnetic wave can propagate through a typical dimension of the system in a time that is shorter than times of interest.

Our capacitive sensor applications are almost all small and slow by these measures, and our conductors are all conductive enough so that their time constant is much shorter than our circuit response times, so we can use these simplified versions of Maxwell's equations

$$\nabla \times \mathbf{E} = -\frac{\partial}{\partial t}\mu_0\mathbf{H} \approx 0$$

$$\nabla \times \mathbf{H} = \frac{\partial}{\partial t}\varepsilon_0\mathbf{E} + \mathbf{J} \approx 0$$

$$\nabla \cdot \varepsilon_0\mathbf{E} = \rho$$

$$\nabla \cdot \mu_0\mathbf{H} = 0$$

A given distribution of charge density ρ produces the electric field intensity $\mathbf{E}$; the magnetic field intensity $\mathbf{H}$ is approximated by zero.

Units

The magnetic permeability of vacuum, μ_0, is a fundamental physical constant, defined in SI units as $4\pi \times 10^{-7}$ N/A^2. The electric permittivity of vacuum, ε_0, is defined by μ_0 and c, the speed of light in vacuum, as $\varepsilon_0 = 1/\mu_0c^2$. As c is defined exactly as 299,792,458 m/s

$$\varepsilon_0 = 8.8541878 \cdot 10^{-12} \quad \text{F/m}$$

Any dielectric material has an electric permittivity which is higher than vacuum; it is measured as the relative dielectric constant ε_r, with ε_r in the range of 2–10 for most dry solid materials and often much higher for liquids.

2.2   CHARGES AND FIELDS

With the simplified equations above, electrostatic analysis reduces to the discovery of the electric field produced by various charge distributions in systems of materials with various dielectric constants.

2.2.1   Coulomb's law

Two small charged conductors in a dielectric with charges of Q_1 and Q_2 coulombs, separated by r meters, exert a force in newtons

$$F = \frac{Q_1Q_2}{4\pi\varepsilon_0\varepsilon_r r^2} \qquad\qquad 2.1$$

The force is along the line connecting the charges and will try to bring the charges together if the sign of their charge is opposite.

The coulomb is a large quantity of charge. It is the charge transported by a 1 A current in 1 s; as an electron has a charge of 1.60206×10^{-19} C, a coulomb is about 6×10^{18} electrons. The force between two 1 C charges spaced at 1 mm is 9×10^{15} N, about 30 times the weight of the earth. But electrostatic forces can often be ignored in practical systems, as the charge is usually very much smaller than a coulomb.

With V volts applied to a parallel plate capacitor of plate area A square meters and spacing d meters, the energy stored in the capacitor is

$$E = \frac{1}{2}CV^2 = \frac{1}{2}\varepsilon_0\varepsilon_r\frac{A}{d} \cdot V^2 \qquad 2.2$$

and the force in newtons is then the partial derivative of energy vs. plate spacing

$$F = \frac{\partial E}{\partial d} = \frac{1}{2}\frac{C}{d}V^2 = \frac{-\varepsilon_0\varepsilon_r}{2}\frac{A}{d^2} \cdot V^2 = -4.427 \cdot 10^{-12}\varepsilon_r\frac{AV^2}{d^2} \qquad 2.3$$

Transverse forces for simple plate geometrys are small, and can be made insignificant with overlapped plates; for some interdigitated structures these forces may be significant and can be calculated using the partial derivative of energy with transverse motion.

For a large air-dielectric capacitor charged to 1 V DC and composed of two 1 m square plates at 1 mm spacing, the force between the plates is attractive at 4.427×10^{-6} N. This force may be troublesome in some sensitive applications. AC operation does not offer a solution to unwanted electrostatic force as both positive and negative half cycles are attractive, but the small force does not affect most capacitive sensor designs, and it can be balanced to zero by use of the preferred three-electrode capacitive sensor geometry. It is exploited in the silicon-based accelerometer of Chapter 15.

Electric field

Two charged conducting plates illustrate the concept of electric field (Figure 2.1).

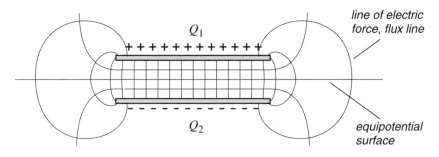

Figure 2.1   Electric field for parallel plates

Voltage gradient

If the plates are arbitrarily assigned a voltage, then a scalar potential V, between 0 and 1 V, can be assigned to the voltage at any point in space. Surfaces where the voltage is the same are equipotential surfaces. The electric field **E**, a vector quantity, is the gradient of the voltage V and is defined as

$$\mathbf{E} = -\frac{dV}{d\mathbf{n}} = -\operatorname{grad} V = -\nabla V \quad 1 \text{ V/m} \tag{2.4}$$

where $\mathbf{n}$ is a differential element perpendicular to the equipotential surface at that point. In the sketch of parallel-plate fields (Figure 2.1), surfaces of constant V are equipotential surfaces and lines in the direction of maximum electric field are lines of force. The voltage along any path between two points a and b can be calculated as

$$V_{ab} = \int_{a}^{b} -\mathbf{E} \cdot d\mathbf{n} \quad V \tag{2.5}$$

In the linear region near the center of the parallel plates (Figure 2.1) the electric field is constant and perpendicular to the plates; eq. 2.5 produces $V_{12} = E\,d$ where d is the plate spacing.

2.2.2   Conduction and displacement current

When an electric field is produced in any material, a current flows. This current is the sum of a conduction current density $\mathbf{J_c}$ and a displacement current density $\mathbf{J_d}$. These terms specify current density in amperes/meter^2.

For a metallic conductor, conduction current is produced by movement of electrons; for electrolytes the current is due to migration of ions. The current density for conduction current is

$$\mathbf{J_c} = \sigma \mathbf{E} \quad \text{A/m}^2 \tag{2.6}$$

where σ is the conductivity in mho/meter.

For a good high resistance dielectric, conduction current is near zero and charge is transferred by a reorientation of polar molecules causing displacement current. Highly polar molecules such as water can transfer more charge than less polar substances or vacuum, and will have a higher dielectric constant ε_r. Also, displacement current is produced by charges accumulating on nearby electrodes under the influence of an applied voltage until the repulsive force of like charges balances the applied voltage. The definition of displacement current is

$$\mathbf{J_d} = \frac{\partial \mathbf{D}}{\partial t} = \frac{\partial}{\partial t}(\varepsilon_0 \varepsilon_r \mathbf{E}) \quad \text{A/m}^2 \tag{2.7}$$

with a direction chosen to be the same as the direction of the $\mathbf{E}$ vector. The displacement current is transient in the case where a DC or a static field is applied, and alternating for AC fields. If a DC voltage is imposed on a system of conductors and dielectrics, displacement current flows briefly to distribute charges until Laplace's equation (eq. 2.10) is satisfied. With an AC field, the alternating displacement current continues to flow with a magnitude proportional to the time rate of change of the electric field.

Gauss' law

The total flow of charge due to displacement current through a surface is found by integrating **D** over that surface

$$\psi = \int_{surf} \mathbf{D}\,ds \quad \text{C} \qquad\qquad 2.8$$

with ds an elementary area and **D** the flux density normal to ds. Gauss' law states that the total displacement or electric flux through any closed surface which encloses charges is equal to the amount of charge enclosed.

The displacement charge ψ is the total of the charges on the electrodes and the charge displaced in the polar molecules of a dielectric in the electric field. Highly polar molecules like water, with a dielectric constant of 80, act as though a positive charge is concentrated at one end of each molecule and a negative charge at the other. As an electric field is imposed, the molecules align themselves with the field, and the sum of the charge displaced during this alignment is the dielectric charge displacement. Note that in a system with finite charges the conduction current density $\mathbf{J_c}$ can be zero for a perfect insulator, but the minimum value of the displacement current density $\mathbf{J_d}$ must be nonzero due to the nonzero dielectric constant of vacuum.

If the excitation voltage is sinusoidal and ε_r and ε_0 are constants, **D** will have a cosine waveform and the displacement current can be found by

$$i = \frac{d\psi}{dt} \quad \text{A}$$

Poisson's equation

The relationship between charge density ρ and displacement current **D** is Poisson's equation

$$\operatorname{div}\mathbf{D} = \rho \quad \text{C/m}^3 \qquad\qquad 2.9$$

div **D**, the divergence of **D**, the net outward flux of **D** per unit volume, is equal to the charge, enclosed by the volume.

Laplace's equation

In a homogenous isotropic medium with ε_r constant and scalar, and with no free charge, Poisson's equation can be rewritten

$$\operatorname{div}\varepsilon_0\mathbf{E} = \varepsilon_0\operatorname{div}\mathbf{E} = \rho = 0 \qquad\qquad 2.10$$

This version of Poisson's equation for charge-free regions is Laplace's equation. This equation is important in electromagnetic field theory. In rectangular coordinates, Laplace's equation is

$$\nabla^2 V = \frac{\partial}{\partial x^2}V + \frac{\partial}{\partial y^2}V + \frac{\partial}{\partial z^2}V = 0 \qquad\qquad 2.11$$

Much of electrostatics is occupied with finding solutions to this equation, or its equivalent in polar or cylindrical coordinates.

> The solution of problems in electrostatics is to find a potential distribution that will satisfy Laplace's equation with given electrode geometry and electrode voltages. Generally, the potential distribution in the interelectrode space and the charge distribution on the electrodes are not known. The charges on the electrodes will distribute themselves so that the conductors become equipotential surfaces and so that Laplace's equation is satisfied in the interelectrode space [Jordan, 1950, p. 48].

A solution of Laplace's (or Poisson's) equation produced the three-dimensional field line plot which was sketched in Figure 2.1. Unfortunately, analytic solution of Laplace's equation is only possible for some simple cases.

Solutions exist for some two-dimensional problems, or for three-dimensional problems which are extruded two-dimensional shapes. Heerens [1986] has published solutions for cylindrical and toroidal electrode configurations with rectangular cross sections which can be extended to many other geometries used in capacitive sensors.

2.2.3   Induced charge

When a positive test charge is brought near a conductor, free electrons in the conductor are attracted to the surface near the charge, and for a floating conductor, holes, or positive charges, are repelled to the opposite surface. With a grounded conductor, the holes flow through the connection to ground and the electrode has a net negative charge. The charges come to an equilibrium in which the repulsive force of the surface electrons is balanced by the attraction of the surface electrons to the test charge.

Electric fields inside a conductor are usually negligible if current flow is small, so the surface of the conductor is an equipotential surface. An electric field outside the conductor but near its surface has equipotential surfaces which parallel the conductor and lines of flux which intersect the conductor at right angles. The magnitude of the conductor's surface charge is equal to the flux density in the adjacent dielectric, $\sigma = |\mathbf{D}|$.

The effect of induced charge is seen in applications such as capacitive proximity detection, as the far-field effects of a capacitive sense electrode must also include the contribution to the E-field of the charge the electrode induces on nearby floating conductors.

2.2.4   Superposition

As for any linear isotropic system, the principle of superposition can be applied to electric field analysis. The electric field of a number of charges can be calculated as the vector sum of the field due to each individual charge. Also, the field in a system of charged conductors can be determined by assuming all conductors are discharged except one, calculating the resultant field and repeating the process with each conductor and calculating the vector sum. Superposition is a very useful and powerful technique for simplifying a complex problem into many simple problems.

2.2.5   Charge images

The distribution of charge on conductors can be determined, often with considerable diffi-culty, by calculating the electric field distribution. Lord Kelvin suggested a simple graphi-cal method. A charge $+q$ in a dielectric near a conducting plane produces a charge density of opposite sign on the nearby surface of the plane. The electric field produced in the dielectric is the same as if the charge density on the surface of the plane was replaced by a single charge $-q$ inside the plane at a symmetrical location. The charge image is similar to an optical image in a mirror (Figure 2.2).

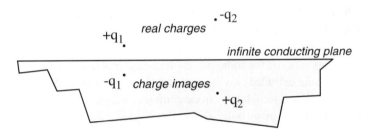

Figure 2.2   Charge images

Lord Kelvin's result can also be derived by looking at the field lines around two charges of opposite polarity and noticing that the line SS is, by symmetry, an equipotential surface which can be replaced by a conducting surface. See Figure 2.3.

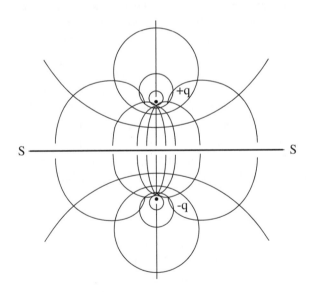

Figure 2.3   Fields of opposite charges

Then no change in the field structure on the $+q$ side of S is seen if the charge at $-q$ is replaced with the induced surface charge on the conductor SS and $-q$ is removed.

Maxwell's capacitor

Maxwell studied a capacitor built with two parallel plates of area *A* and two partially conducting dielectrics of thickness *L*, dielectric constant ε, and conductivity σ (Figure 2.4).

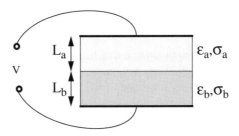

Figure 2.4   Maxwell's capacitor

The terminal behavior and the electric fields can be analyzed by application of Laplace's equation. The diligent student will find that a sudden application of a voltage *V* will produce an electric field which is initially divided between the two regions in proportion to the thickness *L* of the region and its dielectric constant ε, but due to the finite conductivity it will redivide over time in the ratio of the thickness and the conductivity σ. The lazy student will notice that the boundary between dielectrics is an equipotential surface which can be replaced by a conductor, and that the single capacitor can then be dissected into two capacitors with capacitance $\varepsilon_a \varepsilon_0 A/L_a$ and $\varepsilon_b \varepsilon_0 A/L_b$. Then the equivalent circuit is drawn (Figure 2.5)

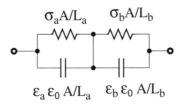

Figure 2.5   Maxwell's capacitor, equivalent circuit

and the circuit is quickly solved by elementary circuit theory, or, for the truly lazy student, SPICE.

2.3   CAPACITANCE

For the parallel plate geometry (Figure 2.1), a voltage *V* can be applied to the plates to produce a total flux ψ. Then, the amount of flux in coulombs which is produced by *V* volts is

$$\psi = Q = CV \qquad\qquad 2.12$$

The new symbol *C* is the capacitance of the plates in coulombs/volt.

Capacitance is calculated by evaluating

$$C = \frac{\varepsilon_0 \varepsilon_r \psi}{\int \mathbf{D} \cdot dl} = \frac{\psi}{\int \mathbf{E} \cdot dl}$$

with dl an elementary length along a flux line of displacement current. This integral gives the capacitance of an elementary volume surrounding the flux line, and must be repeated for all flux lines emanating from one of the plates and terminating in the other plate. For two-electrode systems, all flux lines which emanate from one plate will terminate on the other, but with multiple electrodes this is generally not true.

2.3.1   Calculating Capacitance

Parallel plates

As an example of calculation of capacitance, Gauss' law can be applied to a surface surrounding one of the parallel plates in Figure 2.1. If the surface is correctly chosen and the fringing flux lines at the edge of the plates are ignored, the total charge Q inside the surface is equal to the total displacement flux D times the area of the surface S, resulting in

$$C = \frac{\varepsilon_0 \varepsilon_r S}{d} = 8.854 \times 10^{-12} \times \frac{\varepsilon_r S}{d} \qquad 2.13$$

where

C = capacitance, farads

$\varepsilon_0 = 8.854 \times 10^{-12}$

ε_r = relative dielectric constant, 1 for vacuum

S = area, square meters

d = spacing, meters

Disk

The simplest configuration is a single thin plate with a diameter of d meters. This has a well-defined capacitance to a ground at infinity

$$C = 35.4 \times 10^{-12} \varepsilon_r \times d \qquad 2.14$$

Sphere

$$C = 55.6 \times 10^{-12} \varepsilon_r \times d \qquad\qquad 2.15$$

Two spheres

The capacitance in farads between two spheres of radius a and b meters and separation c is approximately [Walker, p. 83]

$$C \approx \frac{4\pi\varepsilon_0\varepsilon_r}{\dfrac{a+b}{ab} - \dfrac{1}{c}} \qquad\qquad 2.16$$

The approximation is good if a and b are much less than $2c$. Note that with this geometry and the single disk above, capacitance scales directly with size. For the extruded geometries below, capacitance scales directly with the length and is independent of the cross-section size.

Concentric cylinders

Another geometry which results in a flux distribution which can be easily evaluated is two concentric cylinders. The capacitance (farads) between two concentric cylinders of length L and radius a and b meters is

$$C = \frac{2\pi\varepsilon_0\varepsilon_r}{\ln(b/a)}L \qquad\qquad 2.17$$

Parallel cylinders

For cylinders of length L meters and radius a meters separated by b meters, the capacitance in farads is

$$C = \frac{\pi\varepsilon_0\varepsilon_r}{\ln\left(\dfrac{b + \sqrt{b^2 - 4a^2}}{2a}\right)}L \qquad\qquad 2.18$$

If $b \gg a$,

$$C \approx \frac{\pi\varepsilon_0\varepsilon_r}{\ln\dfrac{b}{a}}L \qquad\qquad 2.19$$

Cylinder and plane

A cylinder with length L and radius a located b meters above an infinite plane [Hayt, p. 118] is

$$C = \frac{2\pi\varepsilon_0\varepsilon_r L}{\ln\left[\dfrac{(b + \sqrt{b^2 - a^2})}{a}\right]} = \frac{2\pi\varepsilon_0\varepsilon_r L}{\text{acosh}(b/a)} \qquad 2.20$$

Two cylinders and plane

The mutual capacitance (farads) between two cylinders of length L and radius a meters is reduced by the proximity of a ground plane [Walker, p. 39]

$$C_m \approx \frac{\pi\varepsilon_0\varepsilon_r L \cdot \ln\left[1 + \dfrac{2b}{c}\right]}{\left[\ln\left(\dfrac{2b}{a}\right)\right]^2} \qquad 2.21$$

The approximation is good if $2b \gg a$. A graph (Figure 2.6) of mutual capacitance C_m in pF vs. b in cm, with $L = 1$ m, $a = 0.5$ mm and $c = 2$ cm, and with $c = 20$ cm shows a 20-80× increase of coupling capacitance as the ground plane is moved away from the conductors.

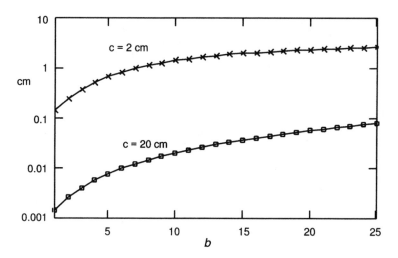

Figure 2.6   Two cylinders and plane, graph of mutual capacitance vs. distance to ground plane

Two strips and plane

A geometry seen in printed circuit boards is two rectangular conductors separated from a grounded plane by a dielectric. For this case, the mutual capacitance in farads between the conductors [Walker, p. 51] of length L, width w, thickness t, and spacing c

meters over an infinite ground substrate covered with a dielectric of thickness t meters with a dielectric constant ε_r is approximately

$$C_m \approx \frac{\pi\varepsilon_0\varepsilon_{r(\text{eff})}L}{\left[\ln\left(\frac{\pi(c-w)}{w+t}+1\right)\right]} \qquad 2.22$$

The effective dielectric constant $\varepsilon_{r(\text{eff})}$ is approximately 1 if $c \gg b$, or if $c \approx b$, $\varepsilon_{r(\text{eff})} \approx (1+\varepsilon_r)/2$. With the strips' different widths w_1 and w_2, the equation becomes [Walker, p. 52]

$$C_m \approx \frac{55.6\varepsilon_{r(\text{eff})}L}{\ln\left[\pi^2 c^2\left(\frac{1}{w_1+t}\right)\left(\frac{1}{w_2+t}\right)\right]} \qquad \text{F, m} \qquad 2.23$$

2.3.2 Multielectrode capacitors

Most discrete capacitors used in electronics are two-terminal devices, while most air-spaced capacitors used for sensors have three or more terminals, with the added electrodes acting as shields or guards to control fringing flux, reduce unwanted stray capacitance, or shield against unwanted pickup of external electric fields.

One use of a three-electrode capacitor (Figure 2.7) is in building accurate reference capacitors of small value [Moon, 1948, pp. 497–507]

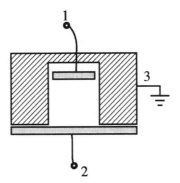

Figure 2.7   Small value reference capacitor

Reference capacitors of this construction were built by the National Bureau of Standards in 1948. A capacitor of 0.0001 pF has an accuracy of 2%, and a capacitor of 0.1 pF has an accuracy of 0.1% [Stout, pp. 288–289]. Electrode 3, the shield, or guard, electrode, acts in this case to shield the sensed electrode (1) from extraneous fields and to divert most of the field lines from the driven electrode (2) so only a small percent of the displacement current reaches the sensed electrode.

In general, the capacitance of a pair of electrodes which are in proximity to other electrodes can be shown, for arbitrary shapes, as in Figure 2.8.

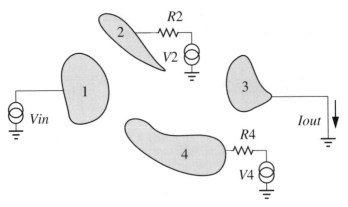

Figure 2.8   Four arbitrary electrodes

The capacitance between, for example, electrode 1 and electrode 3 is defined by calculating or measuring the difference of charge Q produced on electrode 3 by an exciting voltage Vin impressed on electrode 1. If electrodes 2 and 4 are connected to excitation voltages, they will make a contribution to the charge on 3 which is neglected when measuring the capacitance between 1 and 3. The shape of electrodes 2 and 4 and their impedance to ground will have an effect on C_{13}, but the potential is unimportant. Even though a potential change will produce a totally different electric field configuration, the field configuration can be ignored for calculation of capacitance, as the principle of superposition applies; a complete field solution is sufficient but not necessary. If $R2$ and $R4$ are zero and $V2$ and $V4$ are zero, the charge in coulombs can be measured directly by applying a V volt step to Vin and integrating current flow in amps

$$Q_3 = \int Ioutdt$$

Then capacitance in farads is calculated using

$$C_{13} = \frac{Q_3}{Vin}$$

Or, when electrodes 2 and 4 are nonzero

$$C_{13} = \frac{\partial Q_3}{\partial Vin}$$

When $R2$ and $R4$ are a high impedance relative to the capacitive impedances involved, C_{31} is a higher value than if $R2$ and $R4$ are low. Electrodes 2 and 4 act as shields with $R2$ and $R4$ low, intercepting most of the flux between 1 and 3 and returning its current to ground, considerably decreasing C_{13}. With $R2$ and $R4$ high, these electrodes increase C_{13} over the free air value. With linear media, $C_{13} = C_{31}$.

Solving a multiple-electrode system with arbitrary impedances is extraordinarily tedious using the principles of electrostatics. Electrostatic fields are difficult to solve, even approximately. But the problem can often be reduced to an equivalent circuit and handled easily by using approximations and superposition and elementary circuit theory. The four-electrode equivalent circuit, for example, is shown in Figure 2.9.

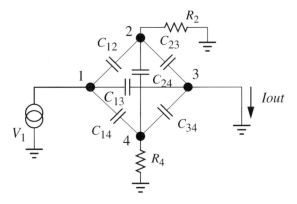

Figure 2.9   Four-electrode circuit

If the time constant of $R2$ with $C12$ and $C23$ and the time constant of $R4$ with $C14$ and $C34$ are small with respect to the excitation frequency, $R2$ and $R4$ may be replaced by short circuits. Then the circuit reduces to that shown in Figure 2.10.

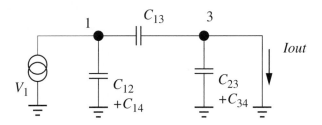

Figure 2.10   Reduced four-electrode circuit

Since a capacitor shunting a low impedance voltage source or a low impedance current measurement can be neglected, the circuit can be further reduced to that shown in Figure 2.11.

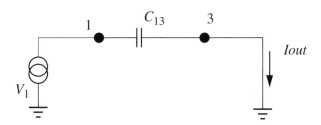

Figure 2.11   Further reduced four-electrode circuit

If possible, it is easier to convert a problem in electrostatics to a problem in circuit theory, where more effective tools are available and SPICE simulations can be used. This type of analysis will be used to understand the effects of guard and shield electrodes with capacitive sensors.

2.4   ANALYTICAL SOLUTIONS

Aside from the easy symmetric cases previously discussed, many other useful electrode configurations have been solved analytically. Some of these solutions are shown in this section.

2.4.1   Effect of Gap Width

Small gaps

For a geometry where adjacent electrodes are separated by a small insulating gap (Figure 2.12)

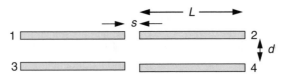

Figure 2.12   Small gap electrodes

Heerens [1986, p. 902] has given an exact solution for the change in capacitance $C' = \delta \cdot C$ between electrodes 1 and 4 or 2 and 3 as a function of gap width

$$\delta = e^{\frac{-\pi s}{d}} \qquad\qquad 2.24$$

Then δ can be plotted against s, with $d = 1$ cm (Figure 2.13).

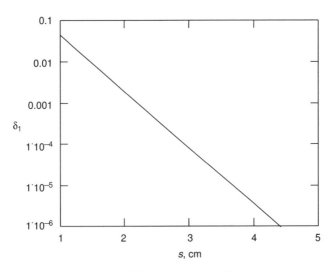

Figure 2.13   Small gap, eq. 2.24

From this plot we see that a physical gap s which is less than 1/5 of the separation d of the electrodes can be considered to be infinitely thin, with an error of less than 10^{-6}. The gap thickness has much less effect on C_{13} and C_{24} than on C_{14} and C_{23}. The electrodes can be considered to have an infinitesimal gap in the center of the actual gap. This rule of thumb, where features less than 1/5 of the plate spacing can be ignored, shows the degree of precision needed to produce accurate capacitive sensors.

2.4.2 Planar Geometries

Overlapping parallel plates

The mutual capacitance of overlapping parallel plates with this geometry (Figure 2.14)

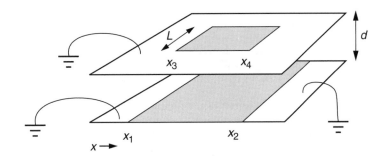

Figure 2.14 Overlapping plates

with unshaded areas separated by a narrow gap and grounded, and the length of the lower electrode infinite, is described [Heerens, 1983, p. 3] as

$$C = \frac{\varepsilon_0 \varepsilon_r L}{\pi} \ln \frac{\cosh\left[\frac{\pi}{2d}(x_4 - x_1)\right]\cosh\left[\frac{\pi}{2d}(x_3 - x_2)\right]}{\cosh\left[\frac{\pi}{2d}(x_3 - x_1)\right]\cosh\left[\frac{\pi}{2d}(x_4 - x_2)\right]} \qquad \text{F, m} \qquad 2.25$$

This equation will be accurate to better than 1 ppm if the length of the lower electrode overlaps the top electrode by more than $5d$ and the gaps between the ground areas and the electrodes are less than 1/5 d. Choosing the following values to illustrate the function, we can evaluate capacitance vs. spacing

$$\varepsilon_0 = 8.854 \times 10^{-12} \qquad x_1 = 0.000$$
$$\varepsilon_r = 1 \qquad x_2 = 0.003$$
$$d = 0 \text{ to } 0.005 \qquad x_3 = 0.001$$
$$L = 0.025 \qquad x_4 = 0.002$$

With these parameters, this curve of capacitance vs. spacing shows an approximately exponential decline which becomes nonlinear near zero spacing (Figure 2.15).

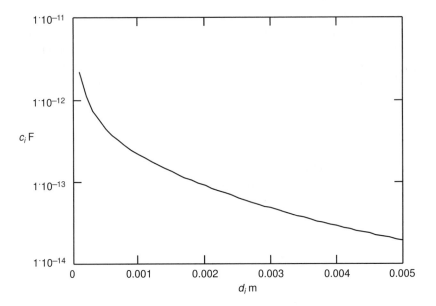

Figure 2.15   Overlapping parallel plate capacitance

Coplanar plates

Two plates in the same plane (Figure 2.16), surrounded again by ground with $L_1 \gg L_2$, have a mutual capacitance [Heerens, 1983, p. 8] which is given by

$$C = \frac{\varepsilon_0 \varepsilon_r L_2}{\pi} \ln \frac{(s + b_1)(s + b_2)}{s(s + b_1 + b_2)} \qquad 2.26$$

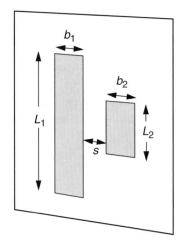

Figure 2.16   Coplanar plates

Coplanar plates with shield

Overlapping parallel plates with the geometry shown in Figure 2.17, with narrow gaps and with the ground planes infinite in extent, or at least five times the d dimension larger than the electrodes, are represented by this equation [Heerens, 1986, p. 901]

$$C = \frac{\varepsilon_0 \varepsilon_r L}{\pi} \ln \frac{\sinh\left[\frac{\pi}{2d}(x_1 - x_3)\right] \sinh\left[\frac{\pi}{2d}(x_2 - x_4)\right]}{\sinh\left[\frac{\pi}{2d}(x_2 - x_3)\right] \sinh\left[\frac{\pi}{2d}(x_1 - x_4)\right]} \qquad 2.27$$

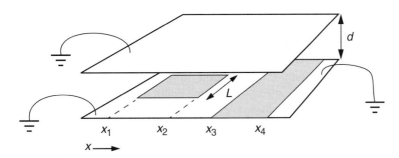

Figure 2.17   Overlapping plates

2.4.3   Cylindrical Geometry

Cocylindrical plates

Two square plates surrounded by ground and mapped to the inside surface of a cylinder [Heerens, 1983] have a capacitance which is independent of the cylinder radius, as with all extruded shapes (Figure 2.18).

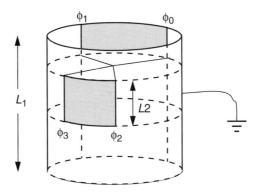

Figure 2.18   Cocylindrical plates

If L_1 is long compared to the length of the shorter plate, L_2, the mutual capacitance of the plates is given by

$$C = \frac{\varepsilon_0 \varepsilon_r L_2}{\pi} \ln \frac{\sin\left(\dfrac{\phi_3 - \phi_0}{2}\right) \sin\left(\dfrac{\phi_2 - \phi_1}{2}\right)}{\sin\left(\dfrac{\phi_2 - \phi_0}{2}\right) \sin\left(\dfrac{\phi_3 - \phi_1}{2}\right)} \qquad 2.28$$

2.5   APPROXIMATE SOLUTIONS

For most capacitive sensor designs, fringe capacitance and stray capacitance can be ignored or approximated without much trouble, but if maximum accuracy is needed, or if problems are encountered with capacitive crosstalk or strays, it is useful to have an analytical method as shown above to evaluate the capacitance of various electrode configurations. Usually it is inconvenient to measure the actual fringe or stray capacitance values, as the strays associated with the measuring equipment are much larger than the strays you are trying to measure. Calculating the strays is possible only for simple geometry with spatial symmetry in a given coordinate system. But an approximate solution is generally adequate; three options that give approximate solutions are field line sketches, Teledeltos™ paper, and finite element analysis.

2.5.1   Sketching field lines

Electric force lines terminate at right angles to conductors. Equipotential surfaces cross the force lines at right angles and tend to parallel conductive surfaces. Starting with a sketch of the conductors and their voltages, Poisson's equation can be solved graphically by trial and error in two dimensions by following these restrictions. Only one set of field lines (or a trivial translation) is produced by a given configuration of conductors. If an additional restriction is followed, that the four-sided shapes formed by the lines are square (or as square as they can get), the field magnitude will be proportional to the closeness of the lines. Field line sketching has been extended to an art by Hayt and others. A simple two-dimensional field sketch is shown in Figure 2.19.

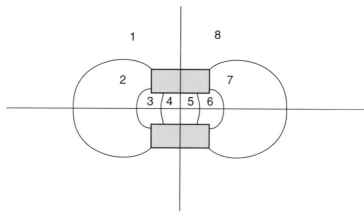

Figure 2.19   Two-dimensional field sketch

Note that some areas like 4 and 5 are nearly square, while areas 1 and 2 are very distorted. The distorted areas can be further subdivided for more precision. After the sketch is finished, block counting is used to estimate capacitance. In the sketch of Figure 2.19 the electrodes are separated by eight blocks sideways (blocks 1–8) and two blocks lengthwise. This has the same capacitance as a parallel-plate capacitor (Figure 2.20) with no fringing fields and a width-to-spacing ratio of 4:1

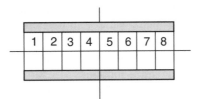

Figure 2.20   Equivalent parallel-plate capacitor

The capacitance is then calculated from eq. 2.14

$$C = \frac{\varepsilon_0 \varepsilon_r S}{d} = 8.854 \times 10^{-12} \times \frac{\varepsilon_r DL}{d} \qquad \text{F, m}$$

where D/d is the electrode width-to-separation ratio and L is the length. With an air-dielectric capacitor having $D/d = 4$, $C = 4 \times 8.854 \times L$ pF, where L is the length in meters.

The example above is a section of a three-dimensional shape which is extruded into the third dimension, and end effects are ignored for calculation of C. Field line sketching techniques can also be extended for nonextruded three-dimensional electrodes with cubical shapes replacing the squares.

2.5.2   Teledeltos paper

Teledeltos[TM] paper is a black resistive paper constructed with a thin carbon coating on ordinary paper backing. It can be used to plot two-dimensional field lines without trial and error. The geometry of the conductors is painted in silver paint and excited by a DC voltage, and a voltmeter is used to determine the equipotential surfaces, or an ohmmeter is used to determine capacitance. Teledeltos paper is also useful to solve for the resistance of two-dimensional shapes such as thin-film resistors used in integrated circuits.

Ordering the supplies

Bob Pease [1994] has contributed instructions on how to obtain this useful but elusive material:

> Simply buy an International Money Order for 44.00 pounds sterling. This pays for everything, including the paper, tax, packing, plus shipping, air freight to anywhere in the USA. (Unfortunately, the fee for the money order will be about $30, but this is an acceptable expense, if you are warned). Send this money order to Mr. David Eatwell at the address [below]. This will soon get you a roll 29 in. wide by 45 ft. long, about 6 kilohms per square, Grade SC20.
>
> Or if you send a money order for 36.50 pounds sterling, you can get a roll 18 in. × 59 ft., tax and air shipping included. Either way, the price per square foot is the

same, about 5 cents per square foot, fairly reasonable, as most experiments take only 1/2 or 1 or 2 square feet.

Mr. David Eatwell, Sensitized Coatings, Bergen Way, North Lynn Industrial Estate, King's Lynn, Norfolk, England PE30 2JL

Your local post office mails the International Money Order and your letter to a U.S. facility in Memphis where it is processed and mailed to England. The normal turnaround time for this service is four to six weeks (the post office does not mail international money orders in Express Mail).

Mr. Pease suggests a two-component silver-loaded epoxy to paint the conductors. One-half oz can be purchased for about $15.00 from Planned Products, 303 Potrero St. Suite 53, Santa Cruz, CA 95060, (408) 459-8088.

Measuring capacitance

After collecting these supplies, the electrode shapes are painted on the Teledeltos paper, and fine copper wires are painted to the electrodes and connected to a voltage source or a resistance meter. The reciprocal of the resistance between electrodes is a measure of capacitance. With paper which has a resistance of 6 kΩ/square, a resistance between two electrodes of 1.5 kΩ would imply four squares in parallel. This is extended to three dimensions by extruding the electrode shapes into the paper, and noticing that the capacitance of parallel plates is proportional to A/d from eq. 2.13. Then the capacitance of the three-dimensional shape is $D/d \times t$, where D/d is the width/spacing ratio and t is the thickness dimension in meters. With four squares in parallel D/d is 4, and with the thickness t meters, the capacitance is $4t \times 0.556 \times 10^{-9} \varepsilon_r$. For the electrode pattern illustrated in Figure 2.21, the capacitance C is

$$C = 0.556 \cdot 10^{-9} \varepsilon_r \cdot \frac{1.5k}{6k} \qquad \text{pF/m}$$

in air, independent of the scale of the electrodes if the cross-section dimensions are scaled together and the thickness dimension is constant.

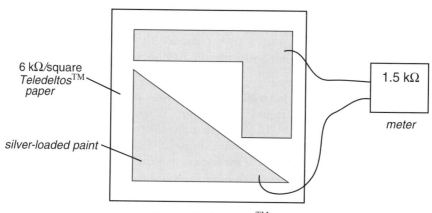

Figure 2.21   Teledeltos^{TM} paper

For field plotting, a 10 V DC supply is connected to the electrodes, and a DC voltmeter with high input impedance (10 MΩ will be fine) is used to plot equipotentials. Multiple electrodes can be simulated, and the shapes can be altered with a pair of scissors as needed to trim designs.

Three-dimensional field plotting has been done using an electrolyte tank. Salt water makes a good electrolyte. Two- and three-dimensional field plotting can also be done by computer, using finite element methods. See the next section, "Finite Element Analysis."

2.6   FINITE ELEMENT ANALYSIS

Since Poisson's law has no direct solution except for some symmetric cases, approximate methods are used. Field line sketching was used by early experimenters, but this is more of an art than a science and it is a cut-and-paste approximation. Teledeltos™ paper (see Section 2.5.2, "Teledeltos paper") offers a more direct solution, but it does not work for three-dimensional problems and it also requires some patience with conductive paint and scissors. Also, due to the tolerance of the paper's resistivity, Teledeltos™ provides a solution accurate only to 5% or so.

A more recent science, finite element analysis, has been used for a variety of problems which can be represented by fields which vary smoothly in an area or volume, and which have no direct solution. FEA was first applied to stress analysis in civil engineering and mechanical engineering, and is now also used for static electric and magnetic field solutions, as well as for dynamic fields and traveling wave solutions.

FEA divides an area into a number of polygons, usually triangles, although squares are sometimes used. Then the field inside a triangle is assumed to be represented by a low-order polynomial, and the coefficients of the polynomial are chosen to match the boundary conditions of the neighbor polygons by a method similar to cubic spline curve fitting or polygon surface rendering. The accuracy of fit is calculated, and in areas where the fit is poorer than a preset constant the polygons are subdivided and the process is repeated. For three-dimensional analysis, the polygons are replaced by cubes or tetrahedrons. A short overview of FEA methods for capacitive sensor design is found in Bonse et al. [1995]. This reference shows FEA error compared to an analytic solution to be less than 0.18%.

FEA is also used by researchers in microwave technology. One approach is to draw a two-dimensional electrode pattern using a shareware geometric drawing package called PATRAN, and pass it to a shareware FEA solver. These programs can be acquired over Internet from an anonymous ftp site, rle-vlsi.mit.edu, at RLE, the Research Laboratory for Electronics at MIT. Ftp "fastcap" from the pub directory.

More convenient FEA software tools integrate drawing and solution packages. MCS/EMAS is available from MacNeal-Schwendler Corp., Los Angeles, CA (213)258-9111. Another, Maxwell, is available from Ansoft, Inc., Four Station Square, Suite 660, Pittsburgh, PA 15219, (412)261-3200. Its features are:

- Integrated modeling, solving, and postprocessing
- Handles electric and magnetic fields
- Solves for fields, energy, forces, capacitances, coupling, etc.
- Runs on Unix workstations, or PCs with Windows or Windows NT

- 2D package about $2500, 3D about $20,000
- 2D package can be configured for *x-y* or *r-θ* coordinate systems
- Parametric analysis option available
- Many different dielectric and conductive materials supported

Maxwell was used to produce the following field charts. The error criterion was set at 0.1%, and the typical solution took 2–15 min on a Pentium 90 processor.

2.6.1   FEA plot

A simple electrode shape demonstrates the steps in FEA analysis. This shape represents a two-dimensional cross section, extruded into the third dimension to a depth of 1 m. The first step is to enter the electrode shape (Figure 2.22).

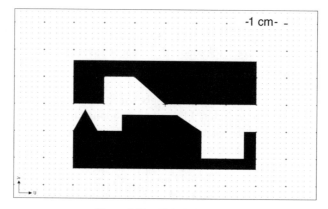

Figure 2.22   FEA electrode shape

Next, the electrodes and the background are assigned material properties. In this case, aluminum was used for electrodes and air for the background. Then the desired error criteria are entered and the project is solved, with the solver adding and subdividing triangles until the requested error bound is reached (Figure 2.23).

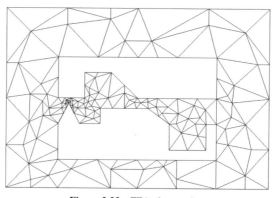

Figure 2.23   FEA plot, mesh

This shows the triangle mesh which was needed to solve the electric field to an accuracy of 0.1%. Usually between 200 and 2000 separate triangles are needed to achieve this level of accuracy. Note the concentration of small triangles near electrode points where the field is changing rapidly.

Equipotentials can then be plotted (Figure 2.24).

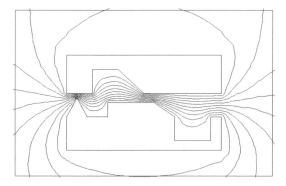

Figure 2.24   FEA equipotential

For this example, the lower electrode was assigned a voltage of 100 V and the upper electrode was assigned 0 V. The equipotentials show the constant-voltage field lines.

With an air dielectric, the interelectrode capacitance was calculated for a 1 m length as 1.003×10^{-10} F. Using eq. 2.13, this capacitance is equal to a 6 cm wide, 1 m long parallel plate capacitor with 0.53 cm spacing.

2.6.2   Fringe fields

FEA plots help to show the effect of fringe fields on the capacity of simple two-plate capacitors. Figure 2.25 shows a thick-plate air-dielectric parallel-plate capacitor, 1 m long with a 1 × 6 cm gap. If the parallel plate formula eq. 2.13 is applied, the calculated capacitance is $\varepsilon_0 \varepsilon_r$ A/s, or $8.854 \times 10^{-12} \times 0.06 / 0.01$ F/m, which evaluates to 53.1 pF.

1 cm

Figure 2.25   Thick plate capacitor, electrodes

The actual capacitance as calculated by FEA is 71.7 pF, 35% larger than the capacitance without considering fringe effects. The absolute capacitance difference due to fringe fields, 18.6 pF, will stay about the same as the gap is decreased or the area of the plates is increased, so it will be a much smaller percentage of the total capacitance for close-spaced geometries.

Beveling the plate edges or using a thinner plate will reduce the fringe capacitance (Figure 2.26).

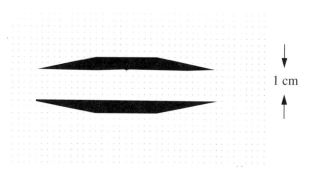

Figure 2.26   Beveled plate capacitor electrodes

Here, FEA calculates 67.5 pF, so the value of the fringe capacitance has decreased to 14.4 pF for 1 m.

Surround with ground

With electrode systems which have a large area relative to the gap, fringe fields will be 1–5% or less and can usually be neglected. If large gaps must be accommodated, surrounding the plates with ground reduces the fringe flux. Surrounding the plates with ground can be done as shown in Figures 2.27 and 2.28.

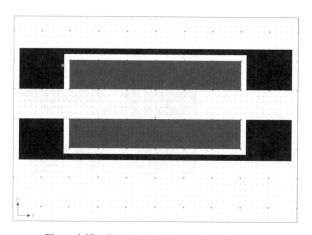

Figure 2.27   Ground shield, electrode configuration

FEA indicates that the capacitance is now reduced to 50.9 pF, indicating a negative fringe capacitance of -2 pF.

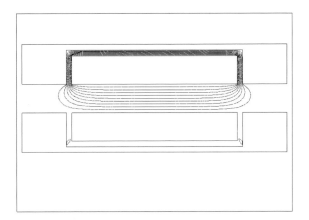

Figure 2.28   Ground shield, equipotentials

2.6.3   Crosstalk

One problem which impairs the performance of two-plate motion sensing capacitors is crosstalk, or unwanted capacitive coupling between, for example, an electrode drive plate and a pickup plate. If this crosstalk is constant it can be canceled in the electronic circuit, but with moving electrodes this is usually not possible. Crosstalk can cause an area-variation motion sensor to falsely indicate transverse motion components; it will give an erroneous indication that a plate has moved transversely when only plate spacing has changed. Luckily, crosstalk diminishes quickly with plate separation, as is shown in this FEA analysis (Figure 2.29).

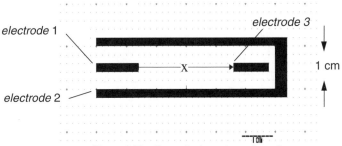

Figure 2.29   Crosstalk electrode configuration

The test signal is applied to electrode 1, and its coupling to electrode 3 is analyzed with electrode 2 grounded. Coupling is defined as the capacitance between 1 and 3 as a percent of the capacitance of 1 to (2 + 3). Parametric analysis shows the variation of coupling with the x dimension increasing from 1 mm to 32 mm (Figure 2.30).

The rapid falloff of coupling with distance is typical. A logarithmic plot shows another typical characteristic, an approximate straight line plot on log-lin coordinates (Figure 2.31).

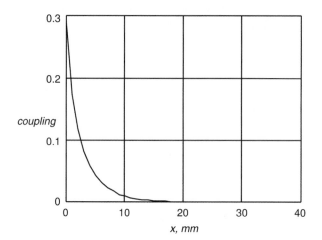

Figure 2.30   Crosstalk, linear parametric plot

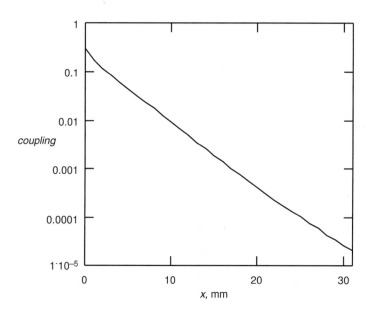

Figure 2.31   Crosstalk, log-linear parametric plot

Another plot, called a line plot (Figure 2.32), shows the variation of electric field along the line marked "x" (Figure 2.29) with the x dimension at 32 mm. Electrode 1 has a test voltage of 100 V applied; electrodes 2 and 3 are grounded.

The crosstalk coupling worsens if the ground does not surround the electrodes. In Figure 2.33 the top shield is removed.

With the top shield missing, crosstalk increases and the curve of coupling vs. distance falls off more slowly with separation. This is shown by adding the one-side curve to the previous two-side shield plot (Figure 2.34).

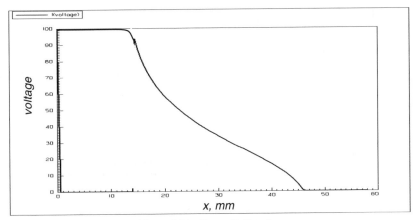

Figure 2.32   Crosstalk, line plot

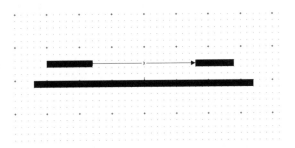

Figure 2.33   Crosstalk, one side shield, electrode configuration

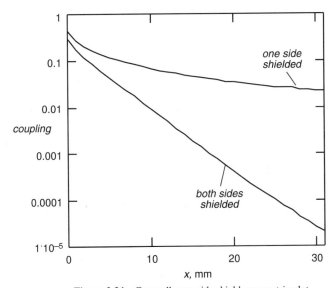

Figure 2.34   Crosstalk, one side shield, parametric plot

2.6.4   Moving shield

If a two-plate moving electrode structure for motion detection is replaced by a three-plate moving-shield structure (Figure 2.35), the capacitance can be approximately first-order insensitive to spacing change. FEA modeling can show how good this approximation is.

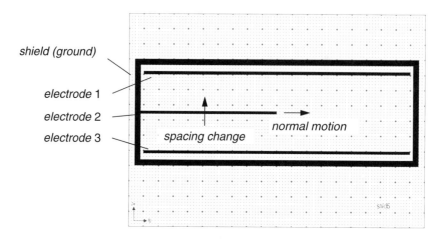

Figure 2.35   Moving shield initial electrode configuration

This represents a cross section of a shielded three-electrode structure used to measure the change of length of electrode 2 by measuring changes in C_{13}. As ground electrode 2 lengthens in the x axis from left to right, the capacitance between electrode 1 and 3 will be linearly decreased. To check how sensitive this structure is to spacing changes, we move electrode 2 in the y axis while leaving its x position unchanged. The final electrode position is shown in Figure 2.36.

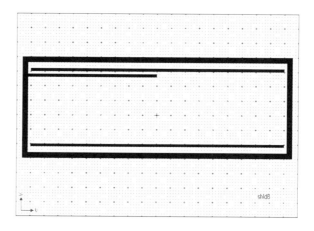

Figure 2.36   Moving shield, final electrode configuration

Figure 2.37 shows that this shape has about 6 mm change in apparent x position as only the electrode spacing is changed. The electrode configuration tested is quite short, however, and has a relatively large gap. If these results are extended to a configuration with a gap which is 1% of the electrode length, the variation of apparent position with spacing will decrease by a factor of 30.

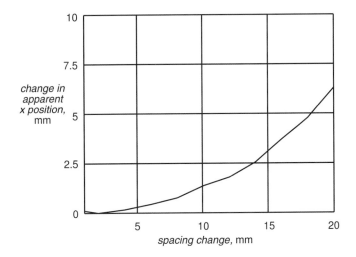

Figure 2.37   Moving shield sensitivity to spacing change

2.6.5   Proximity detector

A possible electrode configuration for proximity detection is analyzed in Figures 2.38–2.40 with FEA. This analysis make use of the r-θ coordinate system as a substitute for true three-dimensional analysis.

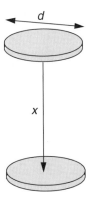

Figure 2.38   Monopole proximity detector, electrode configuration

The change in capacitive coupling between the electrodes with diameter d of 2 cm and a change of spacing x is shown in Figure 2.40.

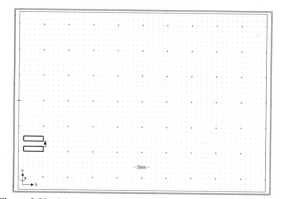

Figure 2.39   Monopole proximity detector, electrode configuration

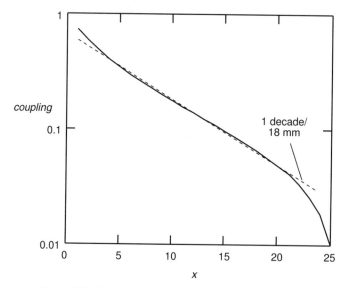

Figure 2.40   Monopole proximity detector, parametric plot, log

As x increases, the distance between electrodes increases and the capacitance falls off as the log of spacing, except as the top electrode approaches the grounded shield the falloff becomes more pronounced. The left boundary of the shielding box is the center of the r-θ coordinate system and does not function as a shield.

3

Capacitive sensor basics

Many of the different types of capacitive sensors use similar plate geometry and similar circuits and components. This chapter presents some of the different electrode configurations used to build various types of sensors.

3.1   BASIC ELECTRODE CONFIGURATION

Three different uses of capacitive sensors are to detect material properties, to sense motion or position, and to detect the proximity of conductive or dielectric objects. These applications all need specialized electrode configurations and circuit design.

The simplest electrode configuration, useful for proximity detection, is a single plate (repeated from Chapter 2) (Figure 3.1).

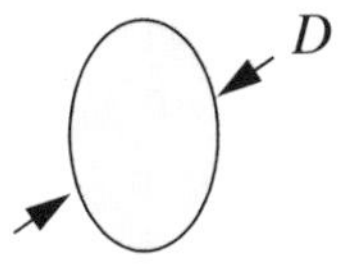

$$C = 0.354D$$

Figure 3.1   Single plate

This assumes vacuum (or air) dielectric. With D in cm, C, the capacitance to an equivalent ground potential very far away, is in pF. A 1 cm plate will have a capacitance of 0.354 pF. As conductive objects approach the plate, this capacitance increases.

The most useful configuration (also repeated from Chapter 2) is two parallel plates (Figure 3.2).

Figure 3.2   Two parallel plates

With dimensions in cm, and an air dielectric, C is in pF. Neglecting fringe effects, the capacitance with 1 cm^2 plate area and 0.1 cm spacing is 0.8 pF. This configuration is useful for measuring dielectric material properties and measuring motion, where either the spacing d is changed or the common area is changed by the transverse motion of one of the plates.

3.2   MOTION SENSING PLATE CONFIGURATIONS

Several more complex plate configurations which optimize linearity or range are used for motion sensing applications.

3.2.1   Spacing variation

For spacing variation motion sensors (Figure 3.3), the capacitance is dependent on the average plate spacing, or more accurately, the integration of $\varepsilon/d \cdot dS$ over the common plate area (assuming no fringing effects), where dS is an elementary area and d is the plate spacing. About 10% change in capacitance is produced by a 10% spacing change, so with small spacing this system is very sensitive to spacing changes. The nonlinear relationship between spacing and capacitance change is a problem if capacitance is measured directly, but the output is linear if capacitive impedance is measured instead.

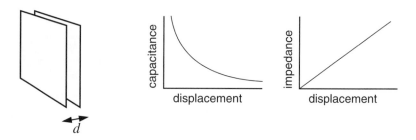

Figure 3.3   Motion sense with spacing variation

This geometry, with both plates equal in size, creates an undesired sensitivity to motion in unwanted axes which is repaired by overlapping or underlapping all edges (Figure 3.4).

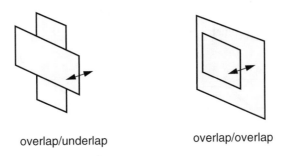

overlap/underlap   overlap/overlap

Figure 3.4   Underlap/overlap

The sensitivity of the system to mechanical disturbance in undesirable axes and to circuit drift can be further improved by using an additional electrode in a bridge circuit (Figure 3.5).

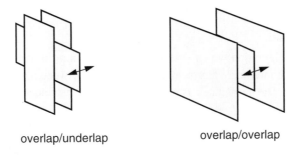

overlap/underlap   overlap/overlap

Figure 3.5   Three-plate bridge

The detecting circuit measures the ratio of the center plate's capacitance to each of the side plates. The advantage of this configuration is that the capacitance can be unaffected by translational motion in the two unwanted axes and can also be relatively insensitive to tilt in any of the three tilt axes. Another advantage of the bridge is that no absolute capacitance reference is needed, and two airgap capacitances in a bridge circuit tend to track quite accurately. The output signal is a ratio rather than an absolute value, and thus can be made to be insensitive to variations of power supply voltage, dielectric constant, and most other circuit and system variables.

A drawback with some amplifier types is the parabolic relationship between motion and capacitance and the limited range of motion which can be measured. The maximum useful linear range for spacing-variation motion detection is usually a small fraction of the plate diameter.

3.2.2   Area variation

For transverse motion which varies area with constant spacing (Figure 3.6), motion and capacitance are linearly related and a long range of motion can be measured, but the sensitivity is less, as 10% change in capacity is produced by a transverse displacement of 10% of the large plate diameter dimension rather than 10% of the small spacing dimension. Area-variation motion sensors can have an unwanted sensitivity to spacing and tilt, especially the two-plate version shown above, but these effects can usually be handled with correct electrode geometry and circuit design.

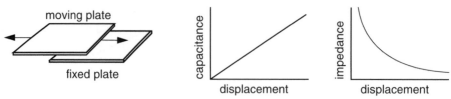

Figure 3.6 Area variation motion sense

Rotary motion

All of the electrode configurations above are shown as linear motion sensors, but they can be easily converted to rotary motion sensors using the familiar x - y to ρ - θ coordinate transform. Rectangular plate shapes will be transformed to pie-shaped sectors.

Moving shield

A configuration which is less sensitive to motion in unwanted axes is the moving shield, which interposes a grounded moving plate between a pair of fixed capacitor plates (Figure 3.7). The capacitance is proportional to the unshielded area of the plates.

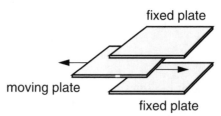

Figure 3.7 Moving shield motion sense

The plate geometries above will almost always need to be protected from external fields with an overall shield or a driven guard electrode, which also serve to control stray capacitance and improve linearity as shown in "Guards and shields" in Section 3.3.3.

3.3   MATCHING THE CIRCUIT TO THE SENSOR

3.3.1   Spacing variation with low-Z amplifier

For the simple parallel-plate spacing sensor used in sensitive capacitive micrometers (Figure 3.8), the output current $Iout$, with A the plate area in meters, x the spacing in meters, vacuum dielectric X_c and the capacitive reactance, in ohms and C, in farads, is

$$I_{out} = \frac{V_I}{X_c} = V_I \omega C = V_I \omega \cdot 8.854 \cdot 10^{-12} \cdot \frac{A}{x} \qquad 3.1$$

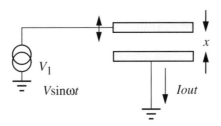

Figure 3.8   Spacing variation, low-Z amplifier

Adding a third plate produces a more stable ratiometric output signal (Figure 3.9).

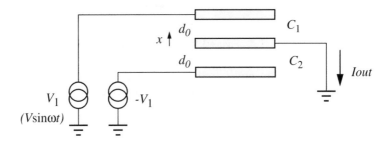

Figure 3.9   Spacing variation, three plates, low-Z amplifier

With the outer plates fixed at a spacing of $2d_0$, when the center plate is moved a distance x the output current is

$$I_{out} = V_I \omega (C_1 - C_2) = V_I \omega \cdot 8.854 \cdot 10^{-12} \cdot A \cdot \frac{2x}{d_0^2 - x^2} \qquad 3.2$$

For small displacements, this becomes

$$I_{out} \approx V_I \omega \cdot 8.854 \cdot 10^{-12} \cdot A \cdot \frac{2x}{d_0^2} \qquad 3.3$$

3.3.2   Spacing variation with high-Z amplifier

If the center plate feeds a high impedance (high-Z) or feedback amplifier, as in Figure 3.13 on page 44, instead of a low impedance amplifier as above, a voltage output is produced. With a voltage output instead of a current output, the response is no longer nonlinear, the electrostatic force is zero, and the dependence on dielectric constant and area disappears

$$V_{out} = 2V_I \left(\frac{X_{C2}}{X_{C1} + X_{C2}} \right) - V_I = \frac{V_I x}{d_0} \qquad 3.4$$

3.3.3   Area variation with low-Z amplifier

A useful area-variation three-electrode geometry for sensing linear displacement is shown in Figure 3.10.

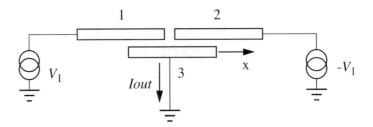

Figure 3.10   Area variation, high-Z amplifier

Here, if electrode 3 is moved to the right a distance x, neglecting fringe effects the inter-electrode capacitance C_{13} will decrease linearly with x while C_{23} increases. $Iout$ is

$$I_{out} = C_{13}\frac{dV_I}{dt} - C_{23}\frac{dV_I}{dt} = (C_{13} - C_{23})\frac{dV_I}{dt}$$ 3.5

and will linearly measure displacement x.

Guards and shields

Fringe fields will corrupt the measurement of position: as the edge of electrode 3 in Figure 3.10 nears the gap between electrodes 1 and 2, the fringe field will make the measurement nonlinear. Also, stray electric fields can be picked up by the signal electrode 3. One fix is to add a guard electrode which surrounds the pickup and moves with it (Figure 3.11).

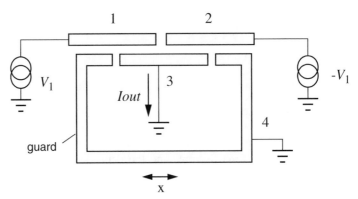

Figure 3.11   Guarded three-electrode displacement sensor

Now, electrode 4 changes the shape of the field lines at the edge of electrode 3 from

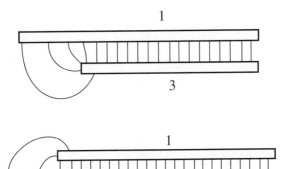

to

The field lines are distorted at the edge of the guard (electrode 4) where this distortion is unimportant, but uniform at the measuring electrode.

The guard electrode also serves as a shield to screen electrode 3 from extraneous fields. Note that the guard electrode does not need to be at the same potential as the signal electrode to be effective [Heerens, pp. 897–898]. It can be at any DC or uncorrelated AC voltage and the same results will be achieved. This is difficult to prove by constructing field lines and applying Poisson's equation, as the field lines will be distorted by the new voltage, but easy to prove using superposition and an equivalent circuit, as in "Multielectrode capacitors" in Section 2.3.1.

3.3.4   Area variation with high-Z amplifier

The previous linear sensor system (Figure 3.11) can be modified to show another way to use the guard electrode (Figure 3.12).

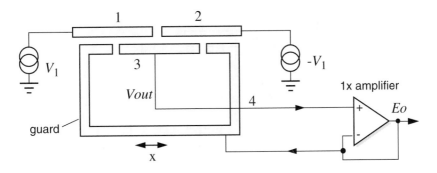

Figure 3.12   Guarded displacement sensor, high impedance amplifier

The output voltage Eo is

$$Eo = V_I \left[\frac{C_{13} - C_{23}}{C_{13} + C_{23}} \right]$$ 3.6

Note that this and many other circuits are shown without a way to control the DC voltage at the operational amplifier input; see Section 4.3.4 for more guidance. Without the guard, the use of a high impedance amplifier would mean that stray capacitance from electrode 3 to local grounds would have attenuated the signal, but with the bootstrapped guard these strays are now unimportant. The previous circuit (Figure 3.11) measured $C_{13} - C_{23}$ by amplifying the output current or output charge, using a low input impedance amplifier. This circuit measures $C_{13} - C_{23}$ by amplifying voltage with a high input impedance amplifier. Each circuit has equivalent performance with respect to rejecting the effects of stray capacitance and rejecting external electric fields, but the circuit of Figure 3.12 has a very important advantage of spacing insensitivity. With the current-output circuit of Figure 3.11, the output with $x = 0$ is zero and is insensitive to spacing, but the maximum output level (and hence the gage factor) is a direct function of spacing. For example, at the maximum-output position with the sense electrode 3 directly opposite electrode 2, the output current is proportional to C_{23}, or inversely proportional to the plate spacing; this is a benefit if the plate spacing is the parameter to be measured, but not for the current task of measuring linear displacement orthogonal to the spacing axis. With the high impedance amplifier circuit (Figure 3.12), the output level is ratiometric rather than absolute response, it is sensitive to V_1 instead of dV_1/dt, and it is totally insensitive to spacing, except for second order effects due to unguarded amplifier input capacitance and fringe fields. The output level is, unfortunately, sensitive to rotation in the axis through the paper, but we will find ways around this in later chapters.

3.3.5   Area variation with feedback amplifier

A third arrangement is similar to the high-Z amplifier above, except the feedback is taken around the driven electrodes (Figure 3.13).

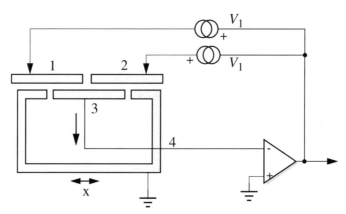

Figure 3.13   Guarded displacement sensor, feedback circuit

The performance of this circuit is very similar to the high impedance amplifier, and it can be made more resistant to stray capacitance. The circuit is similarly insensitive to spacing, as the spacing-variable capacitances C_{13} and C_{23} are inside the feedback loop and in series with the high input impedance of the amplifier. It is discussed further in the next chapter.

3.3.6   Electrode configuration table

Table 3.1 shows the various types of plate and circuit configurations for different applications.

Table 3.1 Capacitive sensor configurations

Application	Plate geometry	Amplifier (Chapter 4)	Circuit	Excitation	See Chap.
Material properties		Low-Z	Single ended or bridge	Variable frequency	10
Micrometer high sensitivity		Low-Z	Bridge	Sine wave	5
Micrometer small movement		Low-Z	Single ended	Square wave 0°/ 180°	5
Motion detect large movement		High-Z or feedback	Single ended	Square wave 0° / 180°	6, 7
Motion detect analog/digital		High-Z or feedback	Sin/cos or tracking	Square wave 0°/ 180°	8
Motion detect 2-axis		High-Z or feedback	Single ended	Square wave 0°/ 90°	19
Proximity detector		Low-Z	Single ended or bridge	Square wave or pseudo-random	6
Silicon sensors		High-Z or feedback	Bridge	Square wave	11, 15

3.4   LIMITS TO PRECISION

3.4.1   Noise

One limit to measurement precision in a capacitive sensor may be the number of charged particles associated with the capacitance. If a test voltage of 1 V is applied to a 1 cm disk, the capacitance of 0.354 pF accumulates a charge $Q = CV$ of 0.354×10^{-12} C. With 6.242×10^{18} electrons/C, we have 2.27×10^6 electrons (or holes) on the disk. In systems where the charges are quantized so that fractional charges are not allowed, as for an isolated electrode, this limits precision to several parts per million. Certainly, the quantization of electron charge will not be a problem for most systems, and fractional charges are usually allowed except for currents through semiconductor junctions or electrodes which are completely floating.

Another limit is the input current noise of the amplifier circuit. With the high impedances of capacitive sensors, current noise is usually more important than voltage noise. With 10 V p-p AC excitation at 100 kHz, the reactance of 0.354 pF is 45 kΩ and the signal current is 10 V/45 kΩ or 2 μA. Bipolar amplifiers with input current noise of 1 pA in a 1 Hz bandwidth would generate 100 pA in a 10 kHz bandwidth and degrade signal to noise ratio to 4000:1. But FET-input operational amplifiers are available with current noise of less than a femtoamp (10^{-15} A) in 1 Hz, so the signal-to-noise ratio measured in a narrow 1 Hz bandwidth can be greater than $20 \times 10^{-6}/10^{-15}$ or 20×10^9. Apparently, current noise is not much of a limiting factor either, except for very small capacitive reactance or large bandwidth.

Actual resolution may be several orders of magnitude worse than this, depending on circuit design; shot noise effects, excess noise in semiconductor junctions, and thermal noise in resistors can be larger than amplifier input noise. But careful design can produce current sensors with noise of just a few thousand noise electrons per second. Systems with a sensitivity of 0.01 aF (0.01×10^{-18} F) have been reported [Jones and Richards, 1973]. Chapter 12 has a more detailed discusson of noise problems with capacitive sensor circuits.

3.4.2   Stability

Environmental

The environmental stability of air-dielectric capacitors is quite good. The dielectric constant of dry air at 0°C and 760 mm pressure is 1.000590, and that of gaseous water (steam) at 110°C is 1.00785, so the effects of temperature and humidity are minimal; Section 12.5 considers these effects in more detail. But absolute capacitance measurement is difficult to do accurately, as stray capacitances must be considered; careful grounding and shielding are needed. If stray capacitances are well guarded, the variation of capacitance due to dielectric constant variation and fringe fields can usually be ignored.

Mechanical

The mechanical stability of the supporting structure can be a concern. For spacing-variation micrometers discussed in Chapter 5, structural stability to μm dimensions is critical, but for area-variation systems or systems with carefully designed reference capacitors, ratiometric bridge amplifiers, and spacing-insensitive circuit design, structural stability is not so much of a problem.

Electrical

With ratiometric circuits, the output is sensitive only to a ratio of capacitances. Very precise components are not needed, and correct circuit design mitigates or cancels the effects of power supply variation and component variation.

4

Circuit basics

The circuit which converts the variable sensor plate capacitance into an output signal should have these characteristics:

- Good linearity
- Shield or guard to isolate the input from stray electric fields
- Insensitivity to stray capacitance to ground on sensor electrodes
- Low noise
- Adequate signal bandwidth
- Correct choice of carrier frequency and waveshape

This chapter discusses the basic circuit building blocks, while Chapter 10 covers these topics in more detail.

4.1   LINEARITY

Capacitive sensors are either area variation, with the plates sliding transversely and no change in spacing as in "Area variation with high-Z amplifier" in Section 3.3.4, or spacing variation where just the spacing changes, as in "Spacing variation with low-Z amplifier" in Section 3.3.1. By looking again at the formula for parallel-plate capacitance

$$C \;=\; \varepsilon_0 \varepsilon_r \cdot \frac{A}{d}$$

and the equation for capacitive impedance

$$Z \;=\; \frac{1}{2\pi f C}$$

we see that *with area-variation sensors, the capacitance is linear with area* but the impedance is not, and *with spacing-variation sensors, the impedance is linear with spacing* but the capacitance is not. Correct circuit design will linearize either type of sensor.

4.2   MEASURING CAPACITANCE

4.2.1   Circuit comparison

The choice of a capacitive sensor circuit (Table 4.1) should consider the required system accuracy, the cost and space available, and the noise environment. At one extreme, a few components only can handle capacitance measurement with the DC circuit, but at a cost of considerable noise sensitivity and drift. For applications needing more precision in a high noise environment, the synchronous bridge circuit is unsurpassed.

Table 4.1 Capacitance measurement circuit comparison: 0 is good, 5 is bad

Circuit	Function	Sensitive to stray capacity	Sensitive to noise	Needs ADC	Bandpass filter available	Size	Sensitive to shunt resistor
DC	volt = $1/C$	yes	5	no	no	0	5
RC oscillator or 1-shot	freq $\cong 1/RC$ period = RC	yes	5	no	no	1	5
IC oscillator (current-capacitance)	freq $\cong 1/RC$ period $\cong RC$	yes	3	no	no	2	5
LC oscillator	freq$\cong 1/\sqrt{LC}$	yes	2	no	yes	3	0
Sync., single ended	volt $\cong C1$-$C2$ or $C1/C2$	no	1	yes	yes	4	1
Sync., bridge	volt $\cong C1$-$C2$ or $1/C1$-$C2$	no	0	yes	yes	5	0

The different capacitance measurement circuits are rated by seven characteristics.

Function

For free-running *RC* oscillators, the output frequency in radians/second is proportional to $1/RC$. The output period is, of course, the reciprocal, or *RC*. With a one-shot *RC* oscillator the ON time varies directly as capacitance.

The two synchronous demodulator circuits can have different functions depending on how the input amplifier is connected; for single-ended circuits the output voltage is a function of a reference capacitor $C1$ and the sense capacitor $C2$.

For synchronous demodulators using a bridge, the output voltage is proportional to $C1 - C2$ or $(C1 - C2)/(C1 + C2)$. In any case, the appropriate function should be chosen to linearize the circuit for spacing-variation or area-variation sensors.

Sensitive to stray capacity

Sensor plates may have signal capacitances in the fractional pF range, and connecting to these plates with 60 pF/m coax would totally obscure the signal. With correct guarding, however, the shield coax and any other stray capacitance can be almost completely nulled out, as shown in Figures 3.11 and 3.12. This guarding is simply a matter of adding a connection with synchronous demodulators, but it is more difficult to guard the oscillator circuits.

Sensitive to noise

One hazard of the oscillator circuits is that the frequency is changed if the capacitor picks up capacitively coupled crosstalk from nearby circuits. The sensitivity of an *RC* oscillator to a coupled narrow noise spike is low at the beginning of a timing cycle but high at the end of a cycle. This time variation of sensitivity leads to beats and aliasing where noise at frequencies which are integral multiples of the oscillator frequency is aliased down to a low frequency. This problem can usually be handled with shields, as shield effectiveness is high for electrostatic fields. Careful power supply decoupling is also needed.

Needs ADC

Frequency is easily converted to a digital number by counting pulses for a fixed time interval, so no ADC may be needed in a typical microcomputer application with *RC* oscillator detectors. Period is similarly converted to digital by counting fixed clock pulses during the measurement interval.

For synchronous demodulators, an ADC may be needed to convert the output voltage to digital, but the ADC can be easily integrated as shown in Chapter 18.

Bandpass filter available

Wideband noise can be substantially reduced for the synchronous demodulators by adding a bandpass filter tuned to the excitation frequency, but no such option is available for the *RC* and *IC* oscillator solutions. This suggests that these oscillators are not the circuit of choice for very sensitive applications.

Circuit size

The oscillators are much lower in component count than are the synchronous demodulators. This is a considerable advantage for conventional printed-circuit-board construction. For integrated circuits, however, switched capacitor methods (see Chapter 11) easily integrate a synchronous demodulator in 1 or 2 mm^2 of area.

Sensitive to shunt resistor

In an application where the circuit may be exposed to contaminants or excessive humidity, resistive paths on the surface of a printed circuit board can affect circuit operation. A very important characteristic of circuit design is sensitivity to resistive or conductive shunts. Correct guarding can handle these problems as well as canceling the effect of stray capacitance, but it is not an available option with many oscillator and one-shot capacitance measurement circuits. An exception is shown in Figure 4.3 on page 52. *LC* oscillators are insensitive to shunt resistance as long as circuit *Q* remains high enough to sustain oscillation, and correctly designed synchronous demodulators are insensitive to shunts.

4.2.2   Direct DC

The first entry in Table 4.1, direct DC, is the simplest detector circuit. With a very high impedance amplifier, capacitance changes can be measured as DC voltage differences, simply by charging the capacitor to be measured and connecting it to the amplifier input. As the charge is then nearly constant the capacitor voltage will vary as the reciprocal of capacitance or directly with spacing by the relationship $Q = CV$. The time constant RC where R is the amplifier input resistance and C the capacitance being measured must be greater than the time measurement period so as not to introduce a low frequency loss; thus an electrometer-type amplifier with input currents in the femtoampere region is often needed. Electret microphones use this method with a junction FET for an amplifier; the resistance to ground which is needed to keep the capacitor voltage from drifting outside the amplifier's linear range is contributed by the FET's input leakage currents.

For direct DC circuits using an operational amplifier an input resistor of very high value is needed, or a bootstrap circuit can be used to increase the AC input resistance, as shown in Figure 4.1.

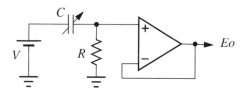

Figure 4.1   DC capacitance circuit

When C varies at frequencies above $1/RC$

$$Eo = \frac{Q}{C} = \frac{C_{avg}V}{C}$$

where Q is the charge on the capacitor C and C_{avg} is the capacitance of C with no displacement.

4.2.3   Oscillator

The direct DC measurement circuit above cannot handle very slow capacitance variations without a very high input impedance amplifier. With a 10 pF capacitor, measuring mV-level signals at 1 Hz requires an amplifier with offset currents in the 0.01 pA range. Also, measurement at DC will admit other unwanted disturbances such as cable noise, thermocouple voltages, power frequency crosstalk, semiconductor $1/f$ noise, and slow variation of component parameters. Circuits which use a high frequency excitation are preferred; for example, the reactive impedance of the unknown capacitor can be measured by using it as the tuning element in an oscillator. Several different types of oscillator can be chosen with different advantages; with an RC oscillator, the frequency is proportional to $1/RC$, but with LC oscillators the frequency is proportional to $1/\sqrt{LC}$ and it is more difficult to linearize. A gyrator circuit (Figure 4.2) which converts capacitance to inductance can be used to save an inductor and to change the output frequency to $1/\sqrt{C1\,C2}$; if $C1$ and $C2$ are both sense electrodes of equal value the response becomes $1/C$.

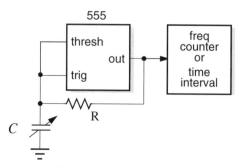

Figure 4.2   *RC* oscillator circuit

Figure 4.2 is a little different from the usual 555 oscillator, but it saves parts, uses a grounded capacitor, produces a 50% duty cycle, and is more linear providing a CMOS-type integrated circuit like the 7555 is used. The normal 555 output does not swing to the power rails and is unstable with temperature in this circuit. The circuit operation is the classic Schmitt-trigger-with-*RC*-feedback, with the Schmitt trigger points accurately determined by the 555 as 1/3 and 2/3 of the power supply. As the output voltage with the CMOS part is accurately driven to the power rails provided that *R* is sufficiently large, the output frequency is unaffected by supply variation. Adding an analog switch allows the measurement capacitor to be switched between several measurement capacitances and a reference capacitor to compensate for changes in resistance and to establish a capacitance-ratio output for greater accuracy.

The output is converted to a digital value using either a frequency counter or a time interval counter with an accurate, usually crystal-controlled, timebase. A frequency counter is appropriate to linearize spacing-variation sensors and a time interval measurement is correct for area-variation sensors.

Guarding the *RC* oscillator

The unguarded *RC* oscillator may have problems with stray capacitance and leakage resistance at the sense capacitor node. The addition of a FET-input op amp as shown in Figure 4.3 produces a low-impedance guard voltage. Adding a guard to the amplifier's *Vee* terminal (Figure 10.9) will further reduce stray capacitance to a fraction of a pF.

Bridge circuit with *RC* oscillator

The advantages of the bridge circuit include a more stable ratiometric response if a matched pair of sense capacitances is used. The *RC* oscillator does not directly accept bridge inputs, but an *RC* oscillator can be configured for a ratiometric response as shown in Figure 4.4.

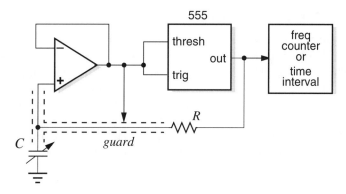

Figure 4.3 Guarded *RC* oscillator

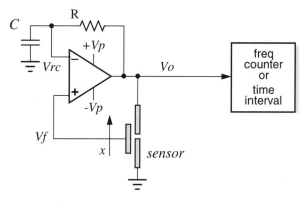

Figure 4.4 Bridge circuit with *RC* oscillator

This circuit shows an *RC* oscillator with the usual output function reversed: frequency is proportional to *RC* instead of 1/*RC*. The op amp output *Vo* is multiplied by a constant *x* which varies linearly between 0 and 1 depending on the sensor plate position, and fed to the positive feedback input, so that *Vf* = *xVo*. With the positive feedback, the op amp output will be either switched to the positive rail +*Vp* or the negative rail -*Vp*. The op amp should be a rail-to-rail output type for good accuracy, and a large value resistor or a switch used on the positive op amp input to set the DC level at that point. The op amp negative input, then, oscillates from +*xVp* to -*xVp* with a time constant of *RC*, and the oscillation period *T* is

$$T = 2RC \cdot \ln\left(\frac{1 + x}{1 - x}\right) \qquad 4.1$$

The circuit is reasonably linear for $x \ll 1$, and can be further linearized by replacing the resistor with a current source of *I* amperes; the equation then becomes linear with *x*

$$T = 2x \cdot \frac{CVp}{I} \qquad 4.2$$

4.2.4   Synchronous demodulator

The most flexible and accurate method of measuring capacitance is to first apply a high frequency signal in the 10 kHz – 1 MHz range through a known impedance to the capacitor under test, then amplify the signal and apply it to a synchronous demodulator. Several variations of the amplifier are available which can appropriately measure either capacitance, C, or impedance, proportional to $1/C$, to produce a linear output, and various input circuit configurations such as bridge or single-ended can be used. With high frequency excitation, electrometer-type very high input impedance amplifiers are not needed, as the capacitive impedance is much lower. Shielding and guarding are easier with a synchronous demodulator than with an oscillator circuit, and a bandpass filter is easily added to limit the noise bandwidth if needed.

The circuit in Figure 4.5 shows a full-wave demodulator, with both positive and negative half-cycles of signal contributing to the output DC level. It can linearize either capacitance- or impedance-variation sensors depending on the configuration of the amplifier.

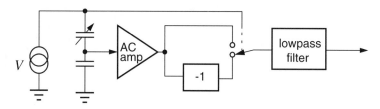

Figure 4.5   Synchronous demodulator circuit

4.3   AMPLIFIERS

With the preferred synchronous demodulator, the amplifier circuit is chosen to produce a linear output for the given sensor configuration, to allow proper guarding and shielding, and to reduce the effect of stray capacitance. Several examples are shown; see also Table 5.1.

4.3.1   High-Z amplifier

The high input impedance amplifier uses a 1× noninverting amplifier configuration as shown in Figure 4.6.

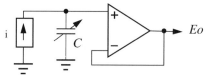

Figure 4.6   High-Z circuit

With an AC current source, this circuit produces an output proportional to the impedance of sensor C, so it produces a linear output with spacing-variation sensors. The input is usually guarded by a shield connected to the output; without this guard, a stray capacitance of only a few pF will seriously attenuate the signal. This stray capacitance is almost completely nulled by a good guard.

4.3.2   Low-Z amplifier

The low input impedance, or virtual ground amplifier (Figure 4.7), will have an input impedance inversely proportional to the gain of the amplifier times the feedback impedance. This connection has similar noise characteristics, better performance with low voltage rails because of its better common mode range, and it can be guarded by use of a grounded shield instead of a floating shield. The output is proportional to $C1 / C2$ or Z_{C2} / Z_{C1} so it can be used to linearize either area or spacing sensors by using the sensor as $C1$ or $C2$.

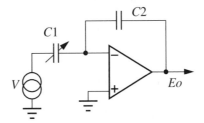

Figure 4.7   Low-Z circuit

4.3.3   Feedback amplifier

The feedback circuit of Figure 3.13 can be redrawn using summing circuits to avoid floating generators and using equivalent variable capacitors to replace sense electrodes (Figure 4.8).

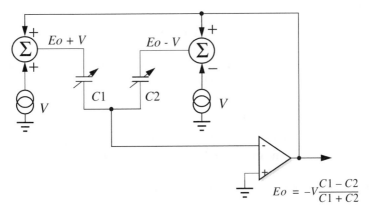

Figure 4.8   Feedback circuit

This circuit has an output proportional to $C1 - C2$ so it has a linear output with area-variation sensors. The input is guarded by ground. This amplifier is particularly good at eliminating the effects of stray capacitance: if we assume we are capacitively coupling to the sense electrodes with a small value capacitor $C3$, the circuit can be redrawn including $C3$ and stray capacitance (Figure 4.9).

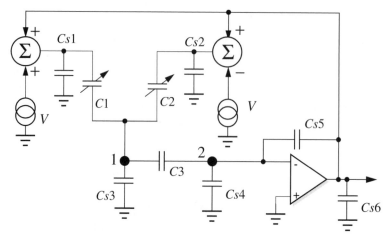

Figure 4.9   Feedback circuit with stray capacitances

$Cs1$, $Cs2$, and $Cs6$ are normally small with respect to the low output impedance of the amplifier and the summers and can be neglected. The circuit of $Cs3$, $C3$, and $Cs4$ will attenuate the signal, but it is inside the amplifier's feedback loop and the only effect the attenuation will have on the circuit is to lower the open loop gain. With a high gain, high frequency amplifier, this will not appreciably affect the closed loop gain. $Cs5$ will directly affect the gain and must be minimized, but this can be handled easily with a two stage amplifier. The feedback circuit has two important advantages over other circuits:

• Insensitivity to stray capacitance
• Guard potential is at ground

In addition, inverting amplifiers are more stable than followers, and amplifier input common mode range, which can be troublesome with followers, is less of a concern. Feedback amplifiers with synchronous demodulators have the best performance and are reccomended for high precision applications.

4.3.4   DC restoration

The circuits of Figures 4.6 – 4.9 will not work for very long, as the amplifier input bias current will cause the output to drift to a power supply rail. This problem can be handled with a high-value impedance across the amplifier input, or across the feedback capacitor for inverting amplifiers, which bleeds off the charge. Some options are:

• Very high value resistor, 100 M or more
• 1 M resistor in T configuration (see Section 10.3.5)
• High impedance FET, or two back-to-back FETs (appropriately high impedance FETs are available only on integrated circuits, not as discrete devices)
• FET switch, momentarily closed at a time when the measurement will not be affected

4.4   SINGLE-ENDED CIRCUITS

Capacitance measurement with a single-ended circuit uses a discrete capacitor, such as an accurate mica or film dielectric component, as a reference to establish the properties of the capacitor under test which is usually an air-dielectric device. One possible circuit is shown in Figure 4.6 where the value of the AC current source is the reference. Another, shown in Figure 4.7, uses a low impedance amplifier and a fixed capacitor as a reference. The output of this circuit is the excitation voltage V times the ratio of the impedance of $C2$ to $C1$. With $C1$ as the variable capacitance, as shown, the output is linear with an area-variation sensor. With $C2$ as the variable and $C1$ as a reference, the output is linear with a spacing-variation sensor. The use of a capacitor as the feedback element rather than the usual resistor improves phase shift and noise performance.

4.5   BRIDGE CIRCUITS

4.5.1   Wheatstone bridge

The standard Wheatstone bridge circuit, shown in Figure 4.10, is often used for low noise instrumentation.

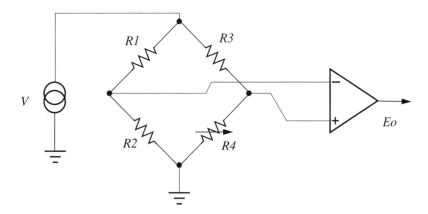

Figure 4.10   Resistor bridge

If the bridge is balanced, $R1/R2 = R3/R4$ and the output is zero. Then large amplifier gain can be used to amplify small differences in one or both legs of the bridge. The advantage of the bridge is that two identical components can be used for $R1$–$R2$ and $R3$–$R4$ so that thermal drifts in these components tend to track and do not affect the balance.

4.5.2   Capacitance bridge

Several variations of capacitance bridges are possible. With a balanced drive replacing the differential amplifier and with capacitances replacing resistors, we have a useful capacitance bridge (Figure 4.11).

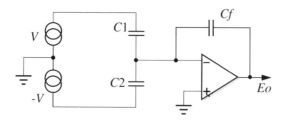

Figure 4.11   Capacitance bridge

$$V_{out} = -V \cdot \frac{C1 - C2}{Cf} \qquad\qquad 4.3$$

The amplifier gain should be high for this equation to be accurate. Any stray capacitance to ground at the amplifier input does not affect the output voltage, but does cut down amplifier available gain at high frequencies, so it should be minimized. $C1$ and $C2$ should be of identical construction for the bridge balance to be stable. The gage factor, or the output voltage divided by input displacement, is a function of $Cf / C1 \parallel C2$, but the null accuracy is a function only of $C1 / C2$. The ratio $C1 / C2$ can be made stable if the capacitors have identical construction.

This type of bridge relies on a balanced drive rather than on good common-mode rejection of the amplifier, and it is generally the preferred circuit. A center-tapped transformer, an inverting amplifier, or a CMOS logic inverter can be used to supply the balanced drive.

4.5.3   Bridge vs. single-ended

The bridge circuit is essential for sensitive circuits which are amplifying a small difference of capacitance, as in Chapter 5, as it is much easier to stabilize a ratio of similar components than a ratio of dissimilar components. The ratiometric output is insensitive to power supply or other circuit variations; with a single-ended circuit these changes would cause a DC error in the output. Also, for circuits with high linear range and low amplifier gain, as in Chapter 7, the bridge is preferred as it can be designed to be insensitive to many circuit parameters. Some additional design details for integrated devices can be found in Kung et al. [1988, 1992].

4.6   EXCITATION

4.6.1   Sine wave

Sine wave excitation is useful for systems needing high frequency (above 1 MHz) carriers, for example, systems which must measure very low capacitance. Sine waves are also preferred for circuits which need high accuracy. Compared to square wave excitation, amplifier slew rate problems are lessened by a factor of 10 and lower frequency amplifiers can be used. Sine wave excitation is essential for high gain bridge circuits, as a good null is more easily achieved without the presence of harmonic energy. But accurate sine wave generation is difficult, and sine wave demodulation uses analog multipliers and other more expensive and less accurate parts compared to square wave circuits.

4.6.2   Square wave

Square wave drives are available virtually free in any system using CMOS logic gates, as the CMOS rail-to-rail output voltage can be quite accurate. Square wave demodulation is generally done with CMOS switches and op amps, and it is easy to integrate for single-chip systems. It is also better suited to low gain systems, such as motion detectors with wide linear operation. For more information, see Section 10.6. For circuits using square wave modulation the designer must be careful to avoid the unstable and nonlinear effects which are produced when an amplifier runs into slew rate limiting, and amplifier band-width must be a factor of ten higher than for sine wave circuits to retain good waveshape.

For both sine and square wave amplifiers, some attention needs to be paid to the amplifier phase shift characteristics. Uncompensated phase shift will typically reduce the demodulator gain.

4.7   FILTERING

4.7.1   Lowpass filter

The circuit of Figure 4.5 includes a lowpass filter to remove demodulation components. The need for this filter can be seen from the spectrum of the signals, shown in Figure 4.12 with sine wave excitation at 20 kHz and motion in the DC-5 kHz frequency range.

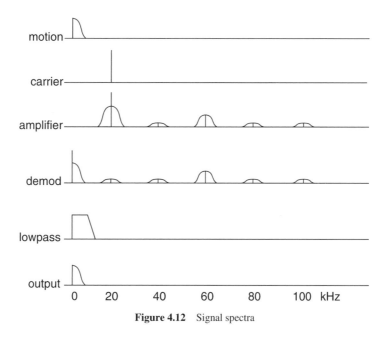

Figure 4.12   Signal spectra

The carrier is assumed to be a sine wave. With changing plate spacing, the amplifier output will be, as shown above, a double sideband replica of the motion signal centered on the 20 kHz carrier, with sidebands produced by the plate spacing variation. Harmonics of 20 kHz are generated by nonlinearities in the amplifier or the carrier or in the multiplication process; often the ±1 circuit shown is used to replace a linear multiplier and the lowpass input is rich with harmonics. With the ±1 circuit, the signal input is linear and the carrier input is hard limited, but the signal linearity and stability can be better than with a conventional two-port linear multiplier.

The demodulator output shows the regenerated motion waveform near DC along with carrier-frequency feedthrough caused by demodulator imbalance, and the upper harmonics generated by the amplifier distortion.

The lowpass filter removes the offending harmonic components and high frequency noise. Also, interfering noise signals at a frequency $\omega 2$ which couple to a sensor excited with $\omega 1$ will produce signals of new frequencies at $\pm(\omega 2 - \omega 1)$, but if $\omega 2 - \omega 1$ is larger than the lowpass bandwidth and the amplifier does not saturate, the lowpass filter output will reject the interference.

4.7.2   Bandpass filter

A further improvement in rejecting noise can be made by inserting a bandpass filter before the amplifier. The amplifier, especially if it has high gain, can generate excessive spurious frequencies with out-of-band noise or slew rate limiting, or in extreme cases it can saturate. These effects are eased with a bandpass filter, centered on the carrier and sized as narrow as possible without affecting the signal bandwidth, preceding the amplifier. If a sharp-cutoff bandpass filter is not used, single-pole highpass and lowpass filters will help.

5
Capacitive micrometers

One of the conceptually simplest and most effective applications of capacitive sensors is for measuring very small spacing changes with close-spaced parallel plane electrodes. Early research dates back nearly to the turn of the century with J. Villey's 1910 publication in *Nature*, and development has continued since, with much research conducted in Europe and the United Kingdom, Jones and Richards describe 15 years of research into ultrasensitive capacitive micrometers, clinometers, gravmeters, and seismographs in their important 1973 paper.

Capacitive micrometers use the capacitance variation caused by changing the plate spacing with two-plate systems, or the capacitance difference or ratio change when the center plate is moved in a three-plate system (Figure 5.1).

Two-plate   Three-plate

Figure 5.1   Two-plate and three-plate micrometers

The capacity of a pair of parallel square plates in air from eq. 2.13, ignoring fringe effects, for a 1×1 cm pair of plates with 0.1 mm spacing, is 8.85 A/d (pF, meters) or 8.85 pF. Edge effects, additional capacitance caused by coupling to the edges or the back of the plates, will increase the capacity by 2–4%. Edge effects are minimized by using close-spaced thin plates, shielding the back, and chamfering the edges. The additional stray capacity due to edge effects is not a strong function of plate spacing and may be assumed constant for most applications.

Electrostatic force for this capacitor (see eq. 2.3) with 10 V excitation is 4.43 μN. Electrostatic effects are normally small, and with the three-plate configuration electrostatic force will be completely nulled when the moving plate is centered. Electrostatic effects are also nulled with a high-Z amplifier.

The design challenge with the ultrasensitive capacitive micrometer is to detect very small variations of plate capacitance and eliminate unwanted environmental variations due to pressure, temperature, and electronic component changes. Researchers have achieved remarkable results [Jones, 1973], with detectable displacement of 10^{-10} mm and a limiting resolution corresponding to a 0.3 aF capacitance change (0.3×10^{-18} farads).

5.1   CIRCUITS

5.1.1   Two-plate micrometer

For less critical applications the RC oscillator can be used to detect plate spacing, but for maximum performance the synchronous demodulator should be used. For the two-plate micrometer (Figure 5.2) the suggested amplifier configuration is a single-ended circuit with a low-Z amplifier. This connection assures that stray capacitance to ground will not affect the performance, and that a grounded shield can be used to shield the device from unwanted local electrostatic fields. This circuit is not useful for very sensitive micrometers as there is no provision to offset the input to amplify small voltages in the presence of DC offset, but it can be used for micrometers with large displacements. DC restoration is not shown; see "DC restoration" (Section 4.3.4).

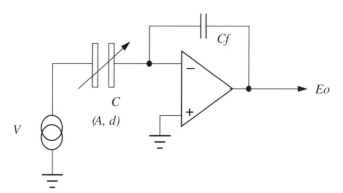

Figure 5.2   Two-plate micrometer

Transfer function of two-plate micrometer

The transfer function of spacing d to output voltage with the low-Z amplifier shown above, with capacitive feedback Cf, is

$$Eo = -\frac{\varepsilon_0 \varepsilon_r A}{Cfd} V \qquad\qquad 5.1$$

which has a parabolic shape. The sensor can be linearized by exchanging it with Cf in the circuit above, or a bridge circuit as in Figure 5.3 may be preferred as its output is higher, it is more linear, and it can be made more stable by matching two bridge capacitors with identical construction.

5.1.2   Three-plate micrometer

The circuit shown in Figure 4.11 is redrawn in Figure 5.3.

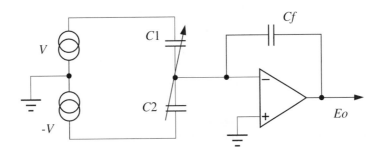

Figure 5.3   Three-plate micrometer

$C1$ and $C2$ represent the capacitances between the micrometer plates, with an average value of C_0 and undeflected spacing d_0. With center plate deflection x, Eo is (Figure 3.9).

$$Eo \;=\; V \cdot \frac{C1 - C2}{Cf} = \; V \cdot \frac{2x}{d_0^2 - x^2} \cdot \frac{C_o}{Cf} \qquad\qquad 5.2$$

Transfer function with low-Z amplifier

The transfer function of the circuit above with $V = 1$ shows a range of linear operation near the center of travel (Figure 5.4).

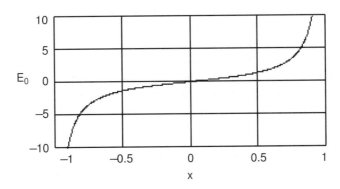

Figure 5.4   Three-plate micrometer transfer function

The input, x, is the normalized center plate position which moves from touching one fixed plate to touching the other. *Eo* is the normalized current output. If the center 20% of this graph is expanded, the curve is more linear (Figure 5.5).

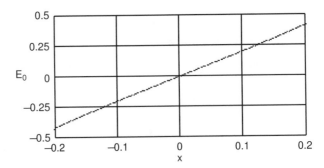

Figure 5.5   Three-plate micrometer transfer function, expanded

A best fit straight line through this curve (Figure 5.6) shows a nonlinearity of less than 1% for almost all of the 20% input range in this expanded scale graph.

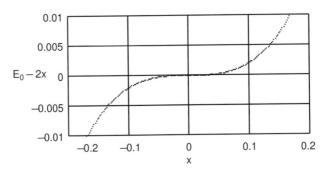

Figure 5.6   Three-plate micrometer transfer function, best fit

The low-Z amplifier is convenient in this application, as a grounded shield can be used to reject extraneous E-fields. If a high-Z amplifier is used, the shield should be connected to the amplifier output. The use of a high-Z amplifier, however, linearizes the bridge. Also, with a high-Z amplifier the electrostatic force on the center plate is zero at all center plate deflections if the deflections are not slower than the amplifier input time constant. The only disadvantage of the high-Z amplifier is the need to use a guard rather than a grounded shield, but with the feedback amplifier circuit (Figure 4.8), the shield may be grounded. The limiting sensitivity of the two types of amplifiers due to noise is similar as shown in Chapter 12. Grounding with the low-Z amplifier is easier than guarding as an extra isolated conductor structure is not needed; a ground shield can usually be fabricated as an extension of an existing conductive enclosure.

The low-Z amplifier is useful for a few applications which can sacrifice linearity for circuit simplicity, but in general it is the last choice for this application due to the extreme nonlinearity.

5.2   CIRCUIT COMPARISON

The different amplifier options for capacitive micrometers can be compared (see Table 5.1)

Table 5.1 Micrometer amplifier comparison

	High-Z (Figure 4.6)	Low-Z (Figure 4.7)	Feedback (Figure 4.8)
Linearity	good	poor, 2% in center 20% of range	good
Shield connected to:	guard voltage	ground	ground
Gage factor depends on:	input cap. to ground	in-out capacitance, Cf	in-out capacitance, summer accuracy
Noise	~ same	~ same	~ same
Parts count	middle	least	most

The high-Z amplifier fixes most of the problems with the low-Z amplifier, but it adds the inconvenience of needing a shield which is not at ground potential, and the gage factor (the gain, or output voltage/input displacement) is affected by the stray input capacitance which often cannot be made zero.

The feedback amplifier uses a grounded shield and the gage factor is sensitive only to amplifier input-to-output capacitance and a resistor ratio. Although a single operational amplifier will usually have an unusably high input-to-output capacitance for small sensors, a two-stage amplifier can be easily designed with negligible input-output capacitance. If the summing circuits are implemented with precision resistors, gage factor is accurate. This circuit is preferred for high precision uses.

6

Proximity detectors

Proximity detectors sense the presence of nearby targets, usually without requiring any contact or wiring to the target or any particular target material properties. Various sensors are available for proximity detection and measurement, including capacitive, inductive, optical, ultrasonic, and magnetic sensors. These detectors are used for many industrial applications and they are available with a variety of analog and digital outputs.

6.1   APPLICATIONS

Typical applications of proximity sensors include [Soloman et al., 1994]:

Motion detection	Detection of rotating motion
	Zero-speed indication
	Speed regulation
Motion control	Shaft travel limiting
	Movement indication
	Valve open/closed
	Conveyer system control
	Transfer lines
	Assembly line control
	Packaging machine control
Process control	Automatic filling
	Product selection
	Machine control
	Fault condition indication
	Broken tool indication

	Product present
	Bottle fill level
	Product count
Sequence control	Verification and counting
	Product selection
	Return loop control
	Product count
Liquid level detection	Tube high-low liquid level
	Overflow limit
	Dry tank
Material level control	Low level limit
	Overflow limit
	Material present

6.2   TECHNOLOGY

Proximity sensors are used as inputs for industrial control systems. Sensors for this application are available with many different outputs to connect to different control systems, including two- and three-wire AC, two-, three-, and four-wire DC, normally open and normally closed switches, TTL logic level, AC/DC, and several four-wire analog output versions including linear 4–20 mA current and 0–5 V voltage output versions. A two-wire analog variable-resistance interface (NAMUR) is often used; these devices are small and easy to wire in a system. NAMUR sensors are calibrated to sink 1.55 mA with a nominal 8.2 V DC supply at the nominal sensing range.

Proximity switches can replace mechanical limit switches for difficult environments. Different versions are used to detect ferrous and nonferrous materials, including glass, cardboard, and plastics. LED lamps are often provided for a visual indication of output state.

6.2.1   Inductive sensors

Inductive sensors generate a local magnetic field at a 2 kHz–500 kHz frequency and detect ferrous materials as an increase in flux for constant excitation, or conductive nonferrous metals are detected as a decrease in flux due to their shielding effect or due to induced lossy eddy currents. Detection range is a function of the size of the magnetic system and is in the range of 1 mm–100 mm; for most industrial sensors the maximum detection range is about the same as the diameter of the magnetic coil. Inductive sensors are rugged, cheap, and reliable, but cannot detect dielectric materials. The range and field topology is similar to capacitive sensors with the limited range similarly due to the inability to focus the magnetic field in air. More power is needed for inductive sensors than capacitive sensors.

6.2.2   Magnetic sensors

Magnetic sensors are similar to inductive sensors, except a DC magnetic field generated by a permanent magnet is used instead of the AC field.

As pickup coils are sensitive only to AC magnetic flux, a Hall effect device which is sensitive to DC magnetic fields is often used. The Hall effect, discovered by Mr. E. F. Hall in 1879, is created in a conductive sheet. With a linear current flowing in one axis, a linear field-dependent voltage is measured in the other axis when a magnetic field is induced through the sheet. Silicon implementations are usually packaged in a plastic transistor case with built-in linear amplifier; output sensitivity is 0.25–2 mV/G with a 25 kHz bandwidth. Some versions are available from vendors such as Allegro Microsystems in Worcester, MA, with Schmitt trigger outputs and high current drivers. A drawback of commercially available Hall sensors is the large and poorly controlled offset voltage which is compensated by the use of AC coupling or computer calibration strategies.

Another magnetic sensor with response to DC uses a saturable high permeability material. Permalloy, for example, has a very high permeability of 100,000 or more, a rectangular hysteresis loop, and low saturation flux density, and can be biased to switch states in response to low level DC magnetic fields.

A switching response is also available with a magnet-actuated reed relay, but sensitivity is not as good.

6.2.3   Optical sensors

Many different types of optical sensors are available, from simple slotted optical gate types to reflective sensors. For very accurate long-range object distance measurement, a coherent laser beam can be bounced off a reflective target, and optical interference effects used to measure position differences by counting interference cycles, with one cycle representing 680 nm for red laser light. At the extreme in complexity, a tunable laser can be used to provide an absolute distance capability as well as the differential distance capability.

Optical sensors can be focused, and they can be very accurate, but often need lenses and special target preparation.

6.2.4   Ultrasonic sensors

Ultrasonic sensors radiate a short ultrasonic pulse in the 20 kHz–500 kHz range. The pulse bounces off a local object and the echo is detected, often by the transducer which launched the pulse. Operation depends on the transmission of air and the sonic reflectivity of the target, which is a function of the orientation and material of its surfaces. Ultrasonics is quite useful in sea water, which attenuates E and H fields but transmits sound well. Soft materials such as cloth and foam do not reflect well.

Sound velocity in air is about 344 m/s, but it is affected by several factors [Eshbach, p. 1092]:

Temperature	+ 0.18%/ °C
Relative humidity	+ 0.4% from 0 to 20% RH
Pressure	little effect below 50 atm
Frequency	- 0.243% for 41 kHz to 1.5 MHz

Some ultrasonic sensors can be adjusted to detect objects only between a preset maximum and minimum range [Turck, p. E4]. As the sound is easily focused, maximum range is higher than inductive and capacitive sensors; maximum range for industrial control units is 6–8 m, while security applications make use of longer range units, with up to 15 m range.

6.3   CAPACITIVE PROXIMITY DETECTORS

Proximity sensing is a simple and effective application of capacitive sensors. Capacitive proximity sensors are used for many applications in plant control, and generally are supplied as a small (1 × 5 cm) cylinder with a pair of electrodes on one end and wire leads on the other. The usual output is a contact closure or TTL-level pulse when an object comes within detection range, about 1 cm. A small amount of hysteresis is added to guarantee dither-free output.

Capacitive proximity sensors are noncontact, can detect small objects, and work with either conducting or insulating objects, such as an unprepared surface of a mechanism or a moving conveyerized object. Similar sized conductive objects are all detected at the same range, but detection of insulating objects depends on a lossy dielectric or a dielectric constant sufficiently different from unity; different insulators are detected at different ranges depending on these parameters and on excitation frequency.

Commercial capacitive proximity sensors specify detection ranges of up to 40 mm for a 34 mm diameter cylinder. One manufacturer [Turck] lists cylinder-style capacitive proximity sensors with diameters between 25 and 80 mm with detection distance approximately proportional to diameter. Several other mechanical configurations are available such as rectangular form factors intended for bolt-in replacements for mechanical limit switches. The output is either a switched AC or DC voltage, analog signals, or a contact closure, as for inductive sensors. Two concentric electrodes are used on the end face, as shown in Figure 6.1.

Figure 6.1   Cylindrical capacitive proximity detector

The active face of capacitive proximity sensors is composed of two concentric metal rings. In one manufacturer's implementation the mutual capacitance between the rings appears as the capacitive feedback element of a high frequency Colpitts oscillator. With no object present, the oscillator is tuned to a point just below oscillation; as a target approaches, the coupling between electrodes increases and the circuit oscillates. The oscillator output feeds a switch or an analog amplitude measurement circuit which generates the digital or analog output voltage.

Dielectric materials are effective at actuating the sensor, with effectiveness increasing with increasing dielectric constant. Conductive materials have a different detection range if they are connected to earth ground; the capacitance of the target to ground produces a decrease in coupling but an increased capacitance to ground. A 40 mm diameter model, the CP40 from Turck, lists these specifications (see Table 6.1).

Table 6.1 CP40 capacitive sensor

Nominal sensing range	20 mm
Max. switching frequency	100 Hz
Input voltage	10–65 VDC
Output contacts	1 NO, 1 NC
Sensing range	
Metal	40 mm
Water	40 mm
PVC	20 mm
Wood (varies with RH)	20–32 mm
Cardboard	8 mm
Glass	12 mm

Turck also lists the dielectric constants shown in Table 6.2.

Table 6.2 Dielectric constants

Material	Dielectric constant	Material	Dielectric constant
Acetone	19.5	Ethylene glycol	38.7
Acrylic resin	2.7–4.5	Fired ash	1.5–1.7
Air	1.000264	Flour	1.5–1.7
Alcohol	25.8	Freon R22, 502	6.11
Ammonia	15–25	Gasoline	2.2
Aniline	6.9	Glass	3.7–10
Aqueous solution	50–80	Glycerine	47
Bakelite	3.6	Hard paper	4.5
Benzene	2.3	Marble	8.0–8.5
Cable sealing compound	2.5	Melamine resin	4.7–10.2
Carbon dioxide	1.000985	Mica	5.7–6.7
Carbon tetrachloride	2.2	Nitrobenzene	36
Celluloid	3.0	Nylon	4–5
Cement powder	4.0	Oil saturated paper	4.0
Cereal	3–5	Paraffin	1.9–2.5
Chlorine liquid	2.0	Paper	1.6–2.6
Ebonite	2.7–2.9	Perspex	3.2–3.5
Epoxy resin	2.5–6.0	Petroleum	2.0–2.2
Ethanol	24	Phenol resin	4–12

Table 6.2 Dielectric constants *Continued*

Material	Dielectric constant	Material	Dielectric constant
Polyacetal	3.6–3.7	Shell lime	1.2
Polyamide	5.0	Silicon varnish	2.8–3.3
Polyester resin	2.8–8.1	Soybean oil	2.9–3.5
Polyethylene	2.3	Styrene resin	2.3–3.4
Polypropylene	2.0–2.3	Sugar	3.0
Polystyrene	3.0	Sulfur	3.4
Polyvinyl chloride resin	2.8–3.1	Teflon	2.0
Porcelain	3.5–4	Toluene	2.3
Powdered milk	3.5–4	Transformer oil	2.2
Pressboard	3.7	Turpentine oil	2.2
Quartz glass	3.7	Urea resin	5-8
Rubber	2.5–3.5	Vaseline	2.2–2.9
Salt	6.0	Water	80
Sand	3–5	Wood, dry	2-7
Shellac	2.5–4.7	Wood, wet	10–30

6.3.1   Limits of proximity detection

Single or dual electrodes

Either a single or a dual electrode structure can be used for proximity detectors. A single electrode produces dielectric flux lines which terminate on the surrounding building structure or on the surface of the earth, and single electrodes can have higher values of detection distance relative to electrode size than dipole electrodes. A hazard is that the earth or building ground potential is available to AC-powered instruments only as the third wire in the power cable, and that connection is usually contaminated by noise, but the use of a high carrier frequency usually successfully avoids this noise component.

The dipole electrode pair is simpler and works as well for low ratios of electrode size to detection distance. Dipole geometry is used in commercially available capacitive proximity sensors.

Earth ground

Capacitive proximity detectors may need to function in environments which are corrupted with extraneous electric fields. One cause of extraneous fields in industrial environments is due to high impedance ground connections.

The "earth ground" in textbooks is a perfectly conducting infinite plane. In real life, earth is not particularly conductive, with a resistivity of 500 Ω-cm or more [Morrison, p. 138]. Sand and gravel can have bulk resistivity of 10 kΩ-cm or more, and a 1 in diameter rod driven to a 10 ft depth can have a resistance of 30 – 50 Ω. Treatment with magnesium sulphate will reduce resistance [Fink et al., pp. 19–52].

Current flow in the earth is due to lightning strikes, power transmission lines, and also earth return currents in power distribution systems. Lightning strikes generate by far the largest earth currents; maximum current can exceed 20,000 A. Power transmission line earth currents are restricted to the area immediately beneath the lines and will not trouble most installations.

Earth current returns from power distribution will be a factor in any industrial situation. The power distribution inside a building is set up so that the "hot" AC connection is made through a black wire, in the U.S., and the return current is nominally through the white wire. Conductive equipment enclosures are earthed through the service neutral, the third (green) wire for three-wire connections. The service neutral is unfused, and it has two functions: to provide a return path for small mA-level leakage currents to the equipment enclosures to return to earth ground for safety, and to provide a return for the much larger fault current which flows if a hot wire is shorted to an enclosure. The fault current then results in a blown fuse instead of a dangerous potential on the enclosure.

Leakage currents to the service neutral result from motor or power transformer insulation leakage and from charging current due to small primary-side capacitors used for EMI control. The service neutral potential in an industrial building relative to earth ground is determined by the sum of these mA-level leakage currents flowing through the service neutral ohmic resistance multiplied by the resistance of the local connection to earth. This potential is usually in the 1–15 V range.

6.3.2   Maximum detection range

It is interesting to calculate the approximate detection range of a carefully designed capacitive proximity detector.

From "Monopole proximity detector, parametric plot, log" (Figure 2.40), the falloff in capacitance with 16 mm diameter electrodes is at a rate of 1 decade / 18 mm of spacing. The capacitance at 1 mm target distance is

$$C_1 = \pi r^2 \cdot \frac{\varepsilon_0}{d} = 1.78 \quad \text{pF} \qquad 6.1$$

For this geometry

$$C = 1.78 \cdot 10^{-d/0.018} \quad \text{pF, m} \qquad 6.2$$

The theoretical limit of detectable capacitance variation from "Limiting displacement of three-plate micrometer," Section 12.2.1, is 3×10^{-22} F for a detection limit of 0.176 m, using a narrow 1 Hz bandwidth. Most proximity detectors have a motion detection bandwidth of 100 Hz; for a 100 Hz bandwidth the theoretic limit degrades to 0.158 m.

The limit of experimentally verified capacitive detection with high voltage (100 V rms) excitation [Jones and Richards, 1973] is 0.05 aF, 0.05×10^{-18} F. Substituting this value for C and solving for d, we have

$$d = 0.018 \log \frac{1.75 \cdot 10^{-12}}{0.05 \cdot 10^{-18}} = 0.136 \quad \text{m} \qquad 6.3$$

That limit is, not unexpectedly, difficult to achieve in practice because of the several effects discussed as follows.

Dielectric constant of air

One limit to the maximum detection distance is due to the change of the dielectric constant ε_r of air. At extreme distances, target movement is indistinguishable from ε_r variations. This section calculates the effect of environmental variations on maximum detection distance.

The dielectric constant of air changes slightly with pressure, temperature, and humidity. At standard temperature and pressure, the dielectric constant changes with temperature as $2 \times 10^{-6}/°C$ for dry air, increasing to $7 \times 10^{-6}/°C$ for moist air. At 20°C, the dielectric constant change with relative humidity is 7×10^{-5} for an RH change from 40 to 90%. A change of pressure of 1 atm changes the dielectric constant by 10^{-4}. Over a distance of a few tens of meters and a time span of a few hours, the expected variation of these parameters might be typically as shown in Table 6.3.

Table 6.3 Change in dielectric constant of air

	Change	Coefficient	Effect
Temperature	5 °C	5 ppm / °C	25 ppm
Relative humidity	10%	1.4 ppm / %RH	14 ppm
Pressure	0.05 atm	100 ppm / atm	5 ppm
Total			44 ppm

The total of the atmospheric variations above will cause the 1.78 pF capacitor of eq. 2.13 to change by 44 ppm, or by 7.8×10^{-5} pF. Using this new value for C and recalculating the limit of detection of the 16 mm circular plate proximity detector (eq. 6.3) we arrive at

$$L = 0.08 \quad m$$

so the maximum detection distance has been cut down by a factor of 20 by these environmental effects. A bridge circuit which uses an airgap capacitor as a reference will compensate for these atmospheric variations, but becomes sensitive to temperature and RH gradients, and constructing an air-spaced capacitor with less than 44 ppm drift is not a simple project.

Local motion

Another limit to the sensitivity of proximity detectors is the effect of small movements of local objects which change the mutual capacitance between the measuring electrodes. This motion may be temperature-induced. The temperature coefficient of aluminum, brass, and steel is in the range of 10–20 ppm/ °C. Depending on the orientation of nearby conductors relative to the measurement electrodes, the effect of a nearby conductor moving only a few μm could considerably decrease detection distance.

Focusing field lines

Sensitivity is also limited by the inability to focus capacitance. The field lines can be controlled locally by use of guarding and shielding electrodes, but once in free space, they spread uncontrollably in response to Poisson's law. Wave-propagated signals like light, sound, or electromagnetic radiation can be accurately focused at considerable distance, but that advantage is not available in capacitive sensors; hence a nearby moving object such as a machine operator may drastically limit the usable maximum distance of a proximity detector. Some improvement in performance is available by positioning grounds appropriately; if the sensor is detecting moving dielectric objects on a conveyer belt, say, the best electrode setup is with the two detection plates on either side of the belt. If that is not possible, with both electrodes on one side, a ground on the other improves field focus by a factor of two or three.

Experimental circuit

An experimental capacitive proximity sensor was built to verify the theoretical limiting range calculated above; it is detailed in Chapter 17.

6.3.3   Proximity sensing equivalent circuit

The proximity sensor electrodes are shown in Figure 6.2.

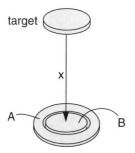

Figure 6.2   Proximity sensor electrodes

A simplified block diagram (Figure 6.3) of the standard two-electrode proximity detector shows the effect of the various capacitances.

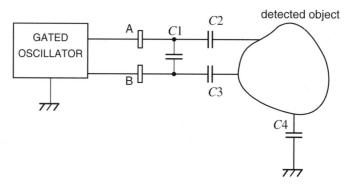

Figure 6.3   Proximity sensor block diagram

The gated oscillator is commonly used for proximity detectors, but it does not have as good a performance as a synchronous detector in noisy environments. The equivalent circuit, for a synchronous detector implementation with electrode A excited, is illustrated in Figure 6.4.

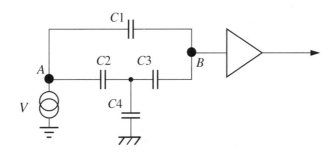

Figure 6.4   Proximity detector equivalent circuit

The synchronous detector and the keyed oscillator circuit can both be analyzed using this equivalent circuit.

The capacitances to be detected, $C2$ and $C3$, are contaminated by stray capacitance $C1$ and $C4$. As target distance increases, $C2$ and $C3$ become much smaller than $C1$, and the additional shunting effect of $C4$ further decreases the signal so that the change in signal with target movement may become a tiny fraction of a percent. Three possible capacitive proximity detector targets may be considered.

Metallic, floating

The equivalent circuit of the floating metallic target is shown in eq. 2.13. $C2$ and $C3$ decrease by about a decade for every sense-plate-diameter increment of target motion.

Metallic, grounded

If the target is a conductive grounded object $C4$ is a short circuit and $C1$ will decrease as the target approaches due to the shielding effect; the change in $C1$ is the measured variable, or a circuit which measures $C2$, the capacitance to ground, can be used.

Dielectric

With a dielectric target $C4$ can be considered open and the values of $C2$ and $C3$ are modulated by the target's proximity. $C1$ increases as a dielectric target approaches due to the increase in permittivity.

6.3.4   Stray fields

Stray electrostatic or electromagnetic fields can interfere with capacitive proximity detection in two ways: by saturating the input amplifier or by being mistaken for the exciting signal.

Input amplifier saturation

An ambient electric field can result from capacitive coupling to a 60 Hz power signal or from the electric field associated with electromagnetic radiation from an RF transmitter. The ambient field may saturate the input amplifier by driving the output to the power rails and the detector will be blocked. Regulatory bodies specify that electronic equipment should not exhibit degraded performance in the presence of a 3 or 10 V/m maximum electric field.

Amplifier saturation is normally not a serious problem, except in very strenuous industrial environments, but it may be avoided by preceding the amplifier with a bandpass filter circuit tuned to the excitation frequency (Figure 6.5).

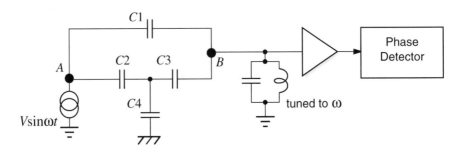

Figure 6.5   Tuned-input proximity detector circuit

Capacitively coupled in-band noise

If, for example, a 20 kHz excitation frequency is used and a local 60 Hz power signal with an irregular waveform is capacitively coupled to the sense electrodes, the 333rd harmonic of 60 Hz is within 20 Hz of the excitation frequency. If the proximity detector's circuits cannot reject this signal, the maximum detection range will be considerably reduced. A resonant circuit or a bandpass filter can help, but a high Q is difficult to achieve and may require tuning. A synchronous detector (below) replaces the bandpass filter with a lowpass filter. This is a very good trade as a 10 Hz lowpass filter is easy to build while a 10 Hz bandpass filter around a 20 kHz carrier is difficult.

Shielding

A metallic shield almost completely attenuates stray E-fields, but proximity detectors cannot usually be completely shielded. Partial shields may help reject local fields.

6.3.5   Phase detectors

Several effective phase detector circuit configurations can be used to virtually eliminate the effects of coupled in-band noise.

Synchronous phase detector

As the detection circuit is in the same enclosure as the transmitter, the detector knows the exact frequency and approximate phase of the signal to be detected. The phase is known exactly if the equivalent circuit is purely capacitive, but in practice resistive elements such as amplifier input impedance will contribute a small phase shift, usually in the range of 10–20°. This phase shift can often be ignored or compensated, and a synchronous detector used (Figure 6.6).

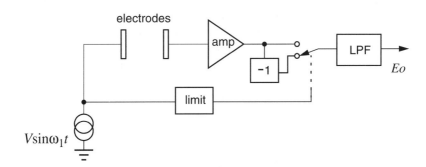

Figure 6.6   Synchronous phase detector

Here, the excitation voltage for the electrodes is also applied to a switch which chooses between the amplified signal electrode voltage and its negative. If phase shifts are small, the result at the LPF output is a DC representation of the magnitude of the electrode voltage at the switching frequency, with switch transients removed by the lowpass filter. The sinusoidal excitation signal, hard limited to a square wave, activates the switch, or a square wave signal feeds both switch and electrodes.

If the phase shifts in the two paths from excitation to multiplier are unequal, the performance degrades until, when the difference in phase shift becomes 90°, the circuit is completely inoperable. Phase shifts can be matched by adding a passive delay line to the control input, usually the shorter path, or by adding a second matched electrode-amplifier with reference capacitors for very accurate tracking.

Pseudorandom carrier

A fix for proximity circuits which need maximum performance in the presence of interfering spectral noise components is to use a pseudorandom excitation voltage, a two-level signal which changes phase in a predictably random way, which is applied to the sense plates and also to the demodulator. The maximal length shift register connection [Rhee, 1989, pp. 265–268] generates a repeating sequence of $2^n - 1$ binary outputs. For example, a 10 bit maximal length shift register outputs $2^{10} - 1$ or 1023 random-appearing binary symbols before repeating. The number of 1s and 0s is almost equal (it differs by 1) and the longest length of a repeated 0 or 1 symbol is $n - 1$. This pseudorandom number (PRN) code can be used directly as a carrier with, say, a 50 kHz clock and a 50 kHz/1023 repetition frequency, and gives good rejection of in-band spectral noise components. With a band-limited amplifier, performance would suffer because of the spectrum-broadening effect of a long run of adjacent 1s or 0s, but a simple digital phase modulator can be added to the PRN output to guarantee a maximum run length of two symbols. The added circuitry for a pseudorandom carrier is minimal; a 10 bit PRN is built with a 10-stage shift register and a quad exclusive-OR gate.

6.4   CAPACITIVE LIMIT SWITCHES

An important type of proximity detector is the limit switch which senses a moving vane and activates a switch closure. For the general case of proximity detector the target cannot be altered, but a limit switch can be designed to use a grounded, dielectric, or floating target, as the target is built as part of the switch. The limit switch should be accurate, inexpensive, and reliable. An example is the optical gate, where a light-emitting diode is paired with a phototransistor to detect any opaque object moved through its 2.5 mm gap.

A capacitive limit switch has several advantages over the optical gate, including much lower power dissipation, lower cost, and more stable operation (Figure 6.7).

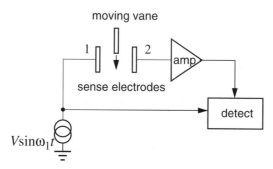

Figure 6.7   Limit switch block diagram

6.4.1   Vanes

The capacitive limit switch detects the difference in capacitive coupling between electrodes 1 and 2 caused by the movement of the vane. The vane may be grounded, and decrease capacitance; floating or dielectric, and increase capacitance; or it may be separately excited.

Grounded vane

To detect a grounded vane, the optimum electrode geometry maximizes coupling when the vane is not present (Figure 6.8).

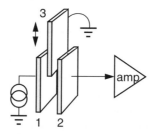

Figure 6.8   Grounded vane electrode geometry

With the grounded vane limit switch, the modulation depth can be very good, especially if the vane is larger than the electrodes; the minimum capacity can be 2–3% of the maximum capacity. The lateral position of the vane does not affect the measurement.

Floating vane

With a floating vane, detection is best done with electrodes which have minimum capacity before the vane is introduced (Figure 6.9).

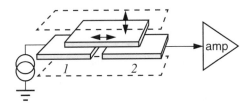

Figure 6.9   Floating vane electrode geometry

Floating vane modulation depth is not as good as grounded vane and, if detecting lateral movement, the vertical position of the vane considerably changes the capacity. Grounded shields installed at the position of the dotted lines will reduce stray capacity and increase modulation depth. These shields will minimize stray capacitance between electrodes 1 and 2, but not affect the coupled capacitance when the moving vane is in position. Motion in the vertical axis can also be detected.

Dielectric vane

The dielectric vane uses an electrode geometry similar to the grounded vane configuration, and the presence of the vane increases electrode capacitance proportional to the vane's dielectric constant and the percent of gap occupied.

Separately excited vane

If the vane can be connected to a high frequency AC source, the performance is improved considerably. In Figure 6.8, for example, this would be done by removing electrode 1 and attaching the exciting signal to electrode 3. Normally this is considered the last choice because of the inconvenience of the extra wire.

Summary of vane performance

A very approximate comparison of the performance of these different limit switch geometries is given in Table 6.4.

Table 6.4 Vane performance

Type	Modulation depth	Sensitivity to gap variation
Grounded vane	97%	low
Floating vane	60%	high
Excited vane	99%	low

6.4.2   Limit switch circuits

The challenge for limit switch circuit design is to reliably detect small changes of capacitance and be insensitive to board contamination and stray resistance or capacitance. For 1 cm square plates and a 2.5 mm gap the capacity is, from eq. 2.13, 0.35 pF, so the detector needs to resolve a change from an open gap at 0.35 pF to a grounded-vane capacitance of 0.05 pF or so. If the electrodes are connected with a 1 cm 0.015 in trace width on 1/16 in glass-epoxy with a ground plane on the opposite side, the PC trace will add a stray capacitance of 0.5 pF. The capacity of an IC pin adds about 4–10 pF to ground. A simple method is needed to detect a fraction of a pF in the presence of much larger strays; the guarding techniques used for the synchronous demodulator circuits would work well, but these circuits are more complex. One method is to use an AC reference.

AC reference

An AC reference, shown in Figure 6.10, is used rather than a DC threshold so that the detection threshold is at ground.

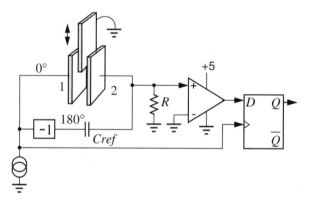

Figure 6.10   Limit switch, AC reference

The waveforms at the excitation terminals and the comparator input are as shown in Figure 6.11.

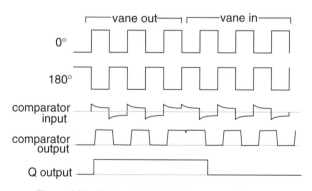

Figure 6.11   Limit switch, AC reference, waveforms

A 5 V 10–200 kHz logic signal can be used for the 0° clock, and a logic inverter generates the 180° clock. If the equivalent capacity between electrodes 1 and 2 varies from 0.05 to 0.35 pF as the vane is moved, *Cref*, the reference capacitor, is chosen to be halfway between the maximum and minimum value of the sense capacitor, or 0.2 pF. *R* is a large-value resistor, over 22 MΩ, and the comparator is a totem-pole-output CMOS type such as Texas Instruments' dual, the TLC3702. The TLC3702 works on a 5 V supply, the input common mode voltage range includes ground, and the input bias current is 5 pA at 25°C temperature.

It is important with a single-supply comparator or amplifier not to exceed the common mode input range with a signal that goes more negative than ground; the usual specification is to limit negative excursions to less than 0.2 V. This is done automatically if stray capacitance is large, as the peak signal is lower than 0.4 V, but if the peak signal is higher than this, a dual power supply or an input bias voltage is needed. Also, the comparator needs to have a minimum delay which is greater than the latch's hold time to avoid a race condition.

As the logic signals can be shared with many switches, the parts to build a switch are just one or two discretes, a comparator, and a latch.

The redeeming virtue of this circuit is that shunt resistance and shunt capacitance strays across the comparator input do not affect the threshold voltage, just the gain. As the comparator has high gain, there is no change in operation with a stray 10 pF or 10 MΩ shunt impedance. If the comparator has nonzero input voltage offset, this is not exactly true; the 3702's 5 mV maximum offset will have a small effect, on detection threshold; adding an AC amplifier fixes it. With 10 pF stray, the signal swing across the comparator input is determined by the capacitive divider to be 5 × 0.2 / 10.2 or about 100 mV, so 5 mV offset change will change the threshold by 5%.

Simpler AC reference

If a microcomputer is available, the circuit can be further simplified. The 0°/180° square wave signals can be replaced by a single interrogation pulse (Figure 6.12).

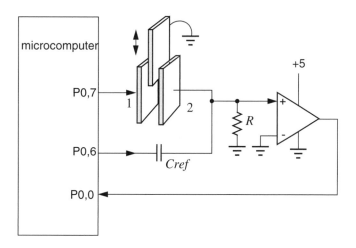

Figure 6.12   Limit switch, single pulse detector

With a microcomputer having an available general-purpose input-output port, this instruction sequence for an 8 bit microcomputer excites and reads the switch.

During power on initialization:

```
out     port0   40H
```

To read the switch:

```
out     port0   80H    ;switch both P0,7 and P0,6
nop                     ;delay by a time interval
nop                     ;longer than the comparator delay
in      port0,0         ;read the comparator; 1= vane out
out     port0   40H    ;reset excitation voltages
```

Noise rejection is handled by multiple reads and discarding a single reading in disagreement with its neighbors.

In Figure 6.13 port 0,7 outputs a positive-going edge at the same time as port 0,6 outputs a negative-going edge. If C_{12}, the coupling capacitance between electrodes 1 and 2, is greater than the reference capacitor, *Cref*, the comparator output will be a zero. *Cref* can be built as a printed circuit board capacitance so its characteristics track with C_{12}. Here, too, stray capacitance across the load resistor does not affect operation if the comparator input offset is small.

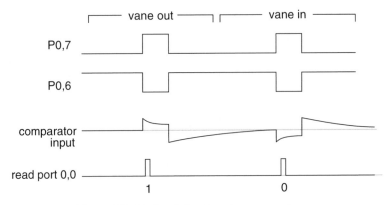

Figure 6.13   Limit switch, pulse reference, waveforms

7

Motion encoders

7.1 APPLICATIONS

Motion encoders are used to convert linear or rotary motion to an analog voltage or a digital code. A variety of motion encoders is available that use many different sensor technologies in a wide size range. Applications include shaft position and speed readout, machine tool control, robotics, servomechanisms, industrial controls, and magnetic media handling. Two basic types are available, absolute and incremental.

7.1.1 Absolute

Absolute encoders output a unique code for each position. An example is an optical rotary encoder which divides 360° of rotation into 1024 parts. A 10 bit output is produced by light sources shining through ten stripe patterns onto ten photosensors. To eliminate the need for accurate edge alignment, the stripe pattern coding is often Gray code rather than straight binary.

7.1.2 Incremental

A 1024-count optical incremental encoder has a single stripe pattern and a single photosensor instead of 10, and calculates absolute position by counting stripes. Absolute position is established by moving the encoder to a limit, or an index output using a second photosensor can be used to reestablish reference position on the fly. A third sensor is used if bidirectional operation is contemplated. Despite the added complexity of establishing absolute position, incremental encoders are considerably simpler than absolute, and have become more popular.

7.2   TECHNOLOGY

Optical stripe encoders

Optical stripe encoders (incremental encoders) are available at low cost if the required resolution is on the order of 100 – 400 counts per turn. Accuracy is excellent, and unpackaged devices are available for incorporation in user-designed mechanics.

Packaged devices with moderate count resolution (512 counts per revolution) are available at less than $10 (MOD5500, BEI Motion Systems, CPD, San Marcos, CA (619)471-2600). As resolution requirements increase, the cost increases to about $30 for a 10,000 count encoder or $65 for a 50,000 count encoder. Maximum stripe densities of over 400,000 counts/revolution are available, but cost over $1000. Also, logic to handle these high counts is expensive: a 20,000 count encoder at 6000 rpm generates 2 MHz clock pulses. This is too fast for direct input to general-purpose microcomputers, so most systems provide a multibit up-down counter to interface to the computer, or use a more expensive DSP computer capable of handling high input frequencies.

A simple radial stripe pattern is usually used. With one sensor, speed can be measured but not direction. With two sensors, direction can be measured also, and if sine and cosine shaping is provided, the distance between stripes can be interpolated for higher resolution. Without sine-cosine shaping, the square wave outputs of two sensors provide double the resolution. These encoders, without interpolation, are used in computer mice, trackballs, and other speed- and position-measuring devices. With analog interpolation for higher resolution, higher effective counts for servomechanisms and accurate position transducers are available.

In closed-loop position servomechanism applications, optical stripe encoders are normally used with either a dedicated PID (proportional-integral-derivative) motion control integrated circuit such as the National LM628, or as an input to a DSP-type computer programmed with a feedback control routine. The DSP (digital signal processing) computer is programmed with the PID algorithm or a fuzzy-logic motion control algorithm.

Binary-coded optical encoders have true absolute position output but are more expensive. A 16 bit (0.0018%) absolute optical encoder costs about $1000 (qty. 1, Teledyne-Gurley, Troy, NY (518)272-6300), a 12 bit (0.0025%) absolute encoder costs $399 (qty. 1, CP-850-12BCD from Computer Optical Products, Chatsworth, CA (818)882-0424).

LVDT

LVDTs, or linear variable differential transformers, are rugged and reliable. They are magnetically coupled and AC excited, capable of accuracy exceeding 0.1%, and are noncontact. The principal drawbacks are the high current consumption and the difficulty of fabricating large magnetic structures; most LVDTs are linear. LVDTs are inherently three-dimensional structures and are difficult to build on a printed circuit substrate.

LVDTs are popular sensors, and an integrated circuit is available which is designed for LVDT interface, Analog Device's AD698.

Resolvers

Resolvers, from vendors such as Litton's Clifton Division, Blacksburg, VA (800)336-2112, look much like an AC motor. A wound rotor is excited by AC and generates AC signals in two stator windings, mechanically disposed at 90° to each other. The stator windings produce sine and cosine waveforms as the rotor is rotated. Higher resolution multipole resolvers produce 16× or 32× cycles per revolution by using many windings and slotted armatures. A typical size 11 (2.5 cm diameter) resolver has an accuracy of ±7 min.

Typical operating temperature range can be –20 to 120°C, and environmental extremes of humidity, shock and vibration, oil mist, coolants, and solvents do not affect operation. Resolvers are rugged and capable of good performance, but no low cost versions are available.

Potentiometers

Linear and rotary motion sensing with positioning accuracy requirements of one part in 1000 can be done with potentiometers. With a precision conductive plastic pot, specs are:

- $10 - 100 \times 10^6$ cycles life
- $0.1°$ backlash
- 0.05% linearity or better (trimmed), 0.005% resolution; 5-10% linearity untrimmed
- High-speed capability to 10,000 rpm
- 0.03% rms noise; about 0.15% p-p
- $3 (10 K) if user provides bearings and packaging; else $5 and up

The main drawback of potentiometers for precision sensor use is the noise caused by the rubbing contact. The quietest available resistance material is conductive plastic, but its noise is many times its resolution, and it will limit performance in high resolution wideband servo systems. The typical high frequency peak characteristic of a PID servo amplifier will cause the output to saturate due to the potentiometer noise before an acceptably high gain can be reached.

Backlash in packaged potentiometers is caused by looseness of fit of bearings and the compliance of the wiper arm working against the coulomb friction at the contact point. When a potentiometer is used as the position sense element in a high gain closed loop position servo, backlash can cause an oscillatory limit cycle behavior when the system is nominally at rest.

Photopot

A "photopotentiometer" replaces the precious-metal wiper used in precision potentiometers with a photoresistor. A precision resistive potentiometer element is coupled to a collector ring using a deposited photoresistive element. The photoresistor is illuminated using a slit aperture which rotates or translates. This device solves the contact noise problems of metal-wiper potentiometers, but at a cost of complexity and power supply current. Linearity is 1.5–3%. Rectilinear models are available (from Silicon Sensors, Inc., Dodgeville, WI, (608)935-2775) to 10 cm length, and rotary models with 2–3 cm diameter.

Summary of position transducers

Table 7.1 compares the position transducers.

Table 7.1 Position transducers

Encoder type	Abs inc	Sin/cos available	Resolution	Supply current	Advantage	Disadvantage
Capacitive	A/I	yes	infinite	0.1 mA	absolute, inexpensive	good mechanics may be needed for accuracy
Hall effect	A	no	infinite	5 mA	low cost	inaccurate
LVDT	A	no	infinite	20 mA	rugged, decent accuracy	rotary not available
Optical stripe	I	yes	50,000 counts	20 mA	accurate, good linearity	incremental, high power, needs counter
Optical absolute	A	no	14 bits	50 mA	absolute, binary output	cost
Photopot	A	no	infinite	20 mA	infinite resolution	one vendor
Potentiometer	A	yes	infinite	0.1 mA	inexpensive	noisy
Resolvers	A	yes	infinite	200 mA	rugged	expensive

Abs/inc

Absolute transducers such as a potentiometer continuously output current position in analog or digital form; incremental transducers like an optical stripe encoder output a pulse each time an increment of position is reached.

Sin/cos

Incremental transcoders output two signals in quadrature which are used for keeping track of direction as well as position. Most versions are digital and use square wave outputs which resolve to half a stripe, but if the outputs are analog and sine and cosine shaped, interpolation between counts can determine a fine position with resolution limited by noise and stability.

7.3   CAPACITIVE MOTION ENCODERS

There are no commercially available modular capacitive motion encoders known to the author, except for capacitive proximity sensors designed for go-nogo proximity detection. But capacitive position sensors are easy to design, the electrode plates can be made as a part of an existing printed circuit board, and no special tooling or hard-to-get parts are needed. These sensors can easily be integrated by the designer on printed circuit boards or integrated circuits and can replace other types of purchased or packaged units.

Good accuracy and resolution are possible with capacitive encoders; one laboratory [de Jong et al., 1994] describes a rotary angle sensor of 50 mm diameter using printed circuit board construction with repeatability of 9 arc seconds, or 17 bits.

7.3.1   Electric potential sensors

Most capacitive position sensors measure the capacity of a system of conducting plates or other shapes. Improved performance for motion detection applications is possible by instead considering the problem as that of measuring electric potential by capacitive coupling. Consider a resistive sheet with an AC voltage imposed and a capacitive plate sensor nearby (Figure 7.1).

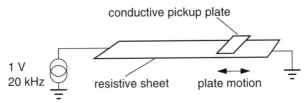

Figure 7.1   Resistive position transducer

Assume the sheet resistance is hundreds of ohms so that the 1 V AC signal produces a linear field just above the surface of the sheet. If the pickup plate is closely spaced, it will pick up the voltage at the surface of the sheet by capacitive coupling. In Figure 7.1, as it moves from left to right, the voltage will change from almost 1 V to zero. The absolute value of the coupling capacitance is relatively unimportant in a properly designed system, as the pickup plate voltage accurately tracks the voltage of the resistive sheet if the pickup is well guarded (Figure 7.25) and if the amplifier input impedance is very high.

This principle can be used to build a class of position sensors which measure fields rather than capacitance. In the case above, in fact, the challenge is to pick up the field capacitively but to eliminate the effects of stray capacitance so that the position measurement is independent of spacing.

Effects of stray capacitance

The model above suffers greatly from stray capacitance. The bottom of the pickup plate is coupled to the resistive sheet with a capacitance of, say, 0.8 pF with a 1 cm^2 area, but the back of the plate will couple to infinity with about 0.2 pF. This reduces the signal by 25% and makes signal amplitude very dependent on spacing. Demodulator designs which are less sensitive to amplitude are possible, but at the cost of a dozen small parts.

If the pickup is in an unshielded enclosure, electric fields generated by radio broadcasts, fluorescent lamps, and 60 Hz wiring will couple easily to the very high impedance pickup amplifier, and stray capacitance to clock-frequency signals will cause a position error with increase of spacing.

Stray capacitance noise pickup can be handled by shielding. As the pickup amplifier is almost totally insensitive to inductive noise, the electric field only must be stopped, and virtually any thickness of conductive material will suffice; see Section 12.3, "Shielding." But capacitance to the back of the pickup plate will be larger if a nearby shield is used, which leads to increased spacing sensitivity; a driven guard is preferable, or a feedback amplifier as shown in Chapter 4.

7.3.2   Alternate ways to generate a linear electric field

Resistive sheet

The resistive sheet in Figure 7.2 could be created by using a screened conductive plastic rectangle on PC board electrodes, with several hundred ohms/square resistivity. This sensor measures pickup displacement in the x direction and can be designed to have good rejection in the y and z directions and in the three tilt axes.

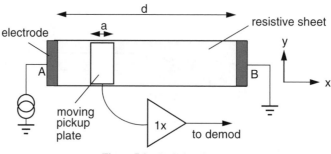

Figure 7.2   Resistive sheet

Note that the pickup is underlapped (or overlapped) so that inadvertent pickup plate displacement in the vertical (y) axis will not affect the detected position. If the pickup is the same width as the resistive sheet, a small displacement in the vertical direction will reduce the pickup area, and if any stray capacitance is present this vertical displacement will reduce the signal and change the reported position. Overlap or underlap by an amount equal to the maximum expected mechanical misalignment plus three to five times the spacing will combat this effect. Tilt around the z axis (into the page) is handled by overlap or underlap. Tilt around the x axis produces no output change, at least first order, and the effect of tilt around the y axis is minimum if $a \ll d$.

Metallic conductors

A metallic conductor pattern can be used as well (Figure 7.3).

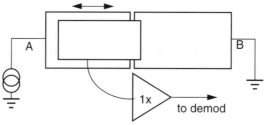

Figure 7.3   Conductive plate, square pattern

This has an advantage over the resistive sheet (Figure 7.2) as the plates can be simply copper areas on a printed circuit board and the pickup area is larger, but a serious disadvantage for accurate sensors as the mechanical sensitivity to pickup plate tilt around the vertical axis is worse due to the increase in pickup plate width. As the right side of the pickup, for example, tilts closer to the substrate because of poor mechanical stability and the left side tilts away, the voltage coupled to the pickup decreases, and the sensor will report a nonexistent change in position.

These plate patterns are drawn as linear transducers, but can easily be applied to a rotary transducer. Wrapping the rectangular plates around a center point produces a rotary pattern reminiscent of air-spaced tuning capacitors (Figure 7.4).

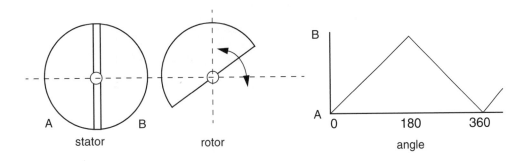

Figure 7.4   Rotary sensor

Adding another similar rotor segment is shown in Figure 7.5.

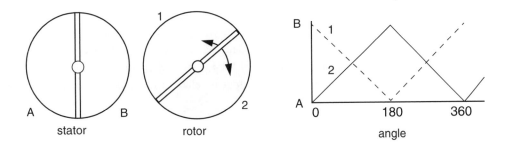

Figure 7.5   Differential rotary sensor

Although needing an additional amplifier, this geometry allows ratiometric detection and is insensitive to spacing, and is insensitive to tilt when the rotor is near 90° intervals, but not when the rotor is near 45°, 135°, etc.

These rotary sensors could have some applications as a rotary transducer, but they have an ambiguity of 180° and are probably not as useful as the sine/cosine plate approach detailed later. A better rotary sensor, with increased linear range, no ambiguity, and with more resistance to tilt, is made by wrapping the V-ramp pattern (Figure 7.8) into a circular pattern.

The ramp pattern shown in Figure 7.6 is equivalent to the resistive sheet (Figure 7.2) if the pickup plate is accurately aligned. It acts as a capacitive divider, so that electrode *A* is coupled tightly to the pickup plate with the pickup at the left end, and *B* is coupled with the pickup at the right end. A high impedance or feedback-type amplifier is preferred to minimize spacing sensitivity, and overlap rather than underlap reduces unwanted sensitivity to *y*-axis displacement.

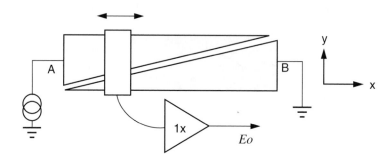

Figure 7.6   Conductive plate, ramp pattern

In the equivalent circuit (Figure 7.7), *C1* is the area common to the pickup plate and the stator section *A*, *C2* is the area common to the stator and *B*. *C1* will increase linearly as the pickup moves to the left, and *C2* will decrease. The voltage at the center tap (the pickup plate) with *Cstray* small is

$$Vp = Vs \times \frac{C1}{C1 + C2} \qquad 7.1$$

or *C1/Ct* where *Ct*, the sum of *C1* and *C2,* is a constant. This equivalent circuit also shows that the capacitive pickup is at least first-order immune to changes in capacitance due to environmental effects: the output is related to a capacitance ratio rather than an absolute capacitance.

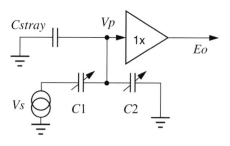

Figure 7.7   Ramp pattern, equivalent circuit

The ramp pattern above (Figure 7.6) suffers from tilt in two axes. Tilt in the vertical (y) axis was discussed and is minimized by using a smaller pickup width. Tilt around the horizontal (x) axis also causes a large change in reported position. The modified ramp pattern (Figure 7.8) shows a simple way to combat this effect.

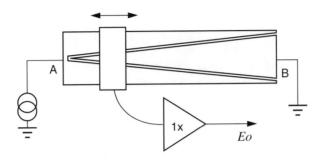

Figure 7.8   V ramp

With this geometry, the plates are symmetric and horizontal-axis tilt is nulled out.

Improved linearity

The V pattern above can be further improved. The simple V pattern produces a non-linear response (Figure 7.9).

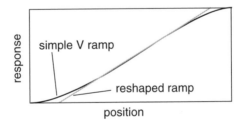

Figure 7.9   Nonlinear response

Reshaping the ramp fixes this (Figure 7.10).

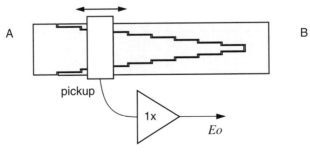

Figure 7.10   Reshaped ramp

The result is a linear relationship between position and output voltage.

With appropriate shaping, the electrodes can generate arbitrary nonlinear functions. With a rectangular pickup as shown, the electrode shape can be determined by an inverse convolution integral as shown in the next section.

7.4   CONVOLUTION

The linear ramp generated by the electrode pattern of Figure 7.10 can be predicted mathematically by the convolution of the pickup shape with the electrode shape. The convolution integral and its inverse can be used to create electrode shapes to generate arbitrary waveforms, with the most useful being linear and sine-cosine shapes for both rectangular and polar coordinates.

7.4.1   The convolution integral

Continuous functions

Convolution is an operation performed on two arbitrary functions, $f(x)$ and $g(x)$. The convolution of these functions, $c(x) = f(x) * g(x)$, is the area under the product of the two functions, plotted against the relative position of the two functions (Figure 7.11).

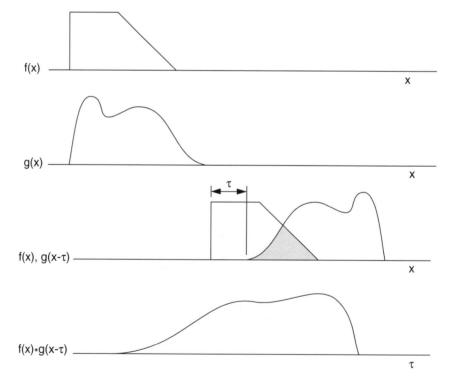

Figure 7.11   Convolution of two functions

The third graph shows $g(x)$ in terms of a dummy variable τ. Since τ is subtracted from x, the shape of $g(x)$ is reversed, and as τ increases from $-\infty$ to $+\infty$, $g(x - \tau)$ moves from right to left relative to $f(x)$. Note that either $f(x)$ or $g(x)$ could be restated in terms of τ, or $\tau/2$ could be added to $f(x)$ and subtracted from $g(x)$ with equivalent results. With τ as shown in graph 3, the region where $f(x) \times g(x)$ is not zero is shown shaded. Using τ as the independent variable the area under $f(x) \times g(x)$ vs. τ is the convolution, as shown in the bottom graph. Its equation is

$$f(n)* g(x) = \int_{-\infty}^{\infty} f(x)g(x - \tau)d\tau \qquad\qquad 7.2$$

Two-dimensional convolution

Convolution is also defined for two-dimensional objects $f(x,y)$ and $g(x,y)$ by calculating the area under $f(x, y) \times g(x - x_1, y - y_1)$ with x_1, y_1 used as dummy variables adjusting the relative displacement of $f(x, y)$ and $g(x, y)$ in two dimensions.

7.4.2   Sampled data convolution

For sampled data systems, the continuous variable x is replaced by the discrete-time variable n and the continuous variable τ is replaced by a discrete-time version of τ, and convolution is then defined as

$$f(n)* g(n) = \sum_{\tau = -\infty}^{\infty} f(n)g(n - \tau) \qquad\qquad 7.3$$

For every finite value of t, all nonzero values of $f(n)$ and $g(n - \tau)$ are multiplied and summed. For mathematics software programs such as Mathcad or MATLAB, with inputs $f(m)$ and $g(n)$ in discrete time (m instead of n to allow independent function lengths), an example calculation (using Mathcad) is

$M:= 100$	define length of $f(m)$
$m:= 0\cdots M{-}1$	define range variable m
$N:= 20$	define length of $g(n)$
$n:= 0\cdots N{-}1$	define range variable n
$f_{(M + n)}:= 0$	set unused $f(m)$ between 100 and 120 to zero
$g_{(N + m)}:= 0$	set unused $g(n)$ between 20 and 120 to zero
$K:= M + N$	set $K = 120$
$k:= 0\cdots K{-}2$	define dummy range variable k
$\tau:= 0\cdots K{-}2$	define range variable τ to span overlap region

$$c_\tau = \sum_{k=1}^{\tau} f_k g_{(K+\tau-k)} \text{convolution } c_\tau \qquad 7.4$$

Here, τ is the variable which moves $g(n)$ relative to $f(m)$; k then scans through all legal values for each new relative position defined by τ. The convolution c_τ multiplies f and g at all values of τ, sums, increments k, and repeats the multiply/sum operation until $k = \tau$. As the input functions are padded with zeros, only values of k in the overlap region are nonzero. This fails in practice, as $K + \tau - k$ can assume values well outside of the range defined for $g(n)$ so that an out-of-range variable will be flagged. Equation 7.4 is repaired by adding truncation to limit range to legal values with a max/min operators, or in Mathcad the mod function

$$c_\tau = \sum_{k=1}^{\tau} f_k g_{\mathrm{mod}(K+\tau-k,\,K)} \qquad 7.5$$

returns the remainder after dividing $K + t - k$ by K and also truncates to legal values.

Another way to calculate the convolution of discrete-time functions is to use the fact that the inverse transform of the product of the complex Fourier transforms of the input signals approximates the convolution.

Inverse convolution

The convolution operation shows the designer of area-variation capacitive sensors what the output signal will look like with predefined electrode shapes. If a particular output signal is needed, this forces a cruel sort of trial and error, exacerbated by the curiously nonintuitive nature of the convolution integral's graphical results. With the inverse convolution operation **iconv**, electrode design becomes more deterministic. If a desired output $c(\tau)$ is defined and one plate $g(x)$ is defined, the other plate $f(x)$ can be calculated by

$$f(x) = \mathbf{iconv}\,[(c(t),\, g(x))] \qquad 7.6$$

This operation may be handled with discrete variables using an element-by-element divide

$$f_m = \frac{\displaystyle\sum_{k=1}^{\tau} f_k g_{\mathrm{mod}(K+\tau-k,\,K)}}{g_0} \qquad 7.7$$

with the obvious restriction that some definitions of f_m and c_τ may not produce a legal g_n. Another obvious restriction is that g_0 should be nonzero; in fact it should be much larger than the limiting precision of the hardware handling the calculations to avoid roundoff errors. Equation 7.6 is used for inverse convolution in Mathcad and was used for the following graphs; another popular mathematics program, MATLAB, has a direct **iconv** operator.

Another way to solve convolution or inverse convolution problems is to use the convolution property of Fourier transforms, where two functions to be convolved are Fourier transformed, multiplied, and inverse Fourier transformed.

7.4.3   Examples of inverse convolution

Equation 7.6 is used to create the following examples (Figure 7.12) of inverse convolution, where one input signal g_n and the convolution c_τ are given and the other input f_m is calculated

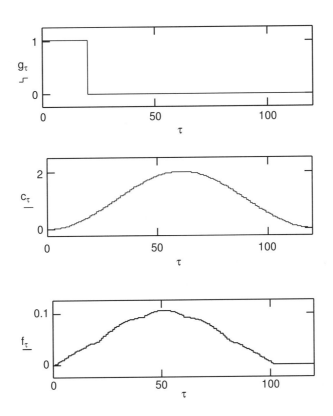

Figure 7.12   Sine wave output

The input g_τ is defined as shown, representing a rectangular pickup plate, and the desired sinusoidal output c_τ is defined as $\sin(n)$. The input function g_τ is graphed vs. range variable τ instead of n so that the full 120 count shape is shown. The inverse convolution equation 7.6 is used to calculate the required shape of the other electrode f_τ, so that if a rectangular pickup plate with width 20 slides linearly over a fixed electrode shaped like f_τ, the output waveform will be the desired sinusoid, c_τ. More examples are shown below (Figures 7.13–7.15)

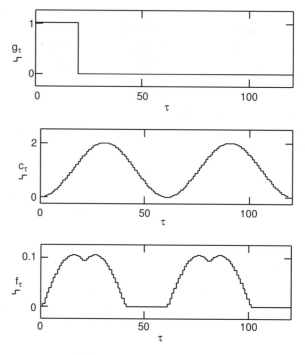

Figure 7.13   Two-cycle sine wave

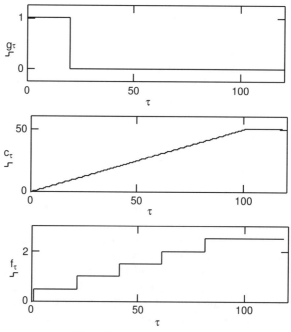

Figure 7.14   Linear ramp output

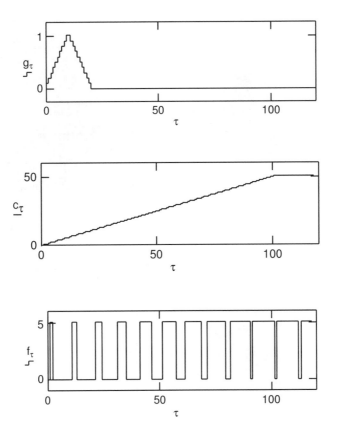

Figure 7.15   Triangular pickup plate

Here, a triangular input and a ramp output are chosen, and the shape of the other input f_τ is calculated. If a nonrectangular pickup is used, as above, the convolution may not yield an exact description of the proper shape of electrodes, as a two-dimensional inverse convolution should properly be used. The example above happens to work, as $f(n)$ is rectangular, but in general it will not and a *y*-axis variable would needed to be added to the calculation as in the two-dimensional convolution examples below.

Two-dimensional convolution

Examples of two-dimensional convolution and deconvolution using Fourier transforms are shown in *The Mathcad Treasury of Methods and Formulas*, 2nd edition, an on-line book of mathematical techniques (see "Convolution, 2D" in the index). With these methods, the convolution of two two-dimensional shapes *A* and *I* by the equation

$$I2 \;=\; \text{icfft}(\text{cfft}(A)(\text{cfft}(I)))$$ 7.8

is shown, using Mathcad's surface graph for display (Figures 7.16–7.19).

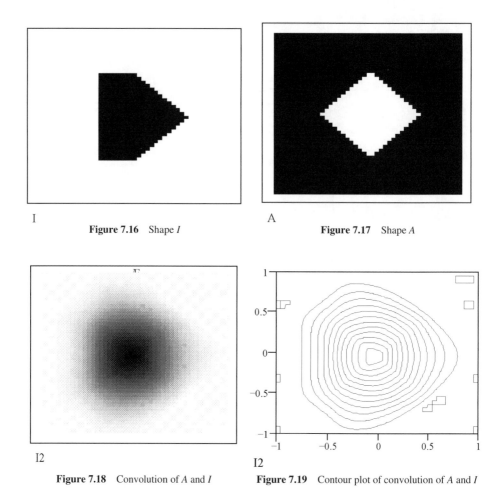

I

Figure 7.16   Shape *I*

A

Figure 7.17   Shape *A*

I2

Figure 7.18   Convolution of *A* and *I*

I2

Figure 7.19   Contour plot of convolution of *A* and *I*

As with the one-dimensional convolution, if the desired output *I2* is known and one input *A* is known, the other input *I* may be derived using deconvolution

$$I \;=\; \mathrm{icfft}\!\left(\frac{\mathrm{cfft}(I2)}{\mathrm{cfft}(A)}\right) \qquad\qquad 7.9$$

The terms icfft and cfft are the inverse complex Fourier transform and the complex Fourier transform. Deconvolution works only if the desired output is realizable by convolution. If only one data point in the desired output is wrong, the mathematics does not converge and a totally bizarre solution is produced.

7.4.4   Rectangular to polar conversion

Any of these rectangular graphs in *x* - *y* coordinates could be converted to *r* - ϕ polar coordinates using the standard transformation

$$r^2 = x^2 + y^2$$

$$\tan\phi = \frac{y}{x}$$

A rectangular plate transforms to a pie-shaped plate in polar coordinates (Figure 7.20).

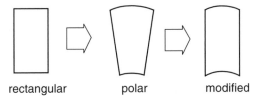

rectangular   polar   modified

Figure 7.20   Pickup transformation

Note that the pickup plate can be pie shaped, as the polar coordinate transformation would suggest, but sensitivity to runout is smaller with the modified rectangular pickup shape.

7.5   PLATE GEOMETRY

The stray capacitance situation is improved if the pickup plate is enclosed by the stator plates. When the pickup moves away from one stator, suffering a loss of capacitive coupling, it moves closer to the opposite plate. The increase from one plate compensates for the decrease from the other, if the pickup is approximately centered. The pickup can be completely enclosed by the stator, either in the cylindrical geometry shown in Figure 7.21 or using two stator plates (Figure 7.22).

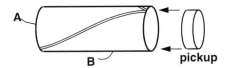

Figure 7.21   Cylindrical plates

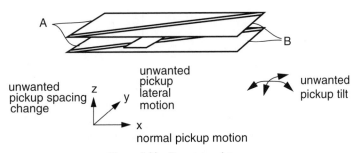

Figure 7.22   Two stator plates

Another benefit is 2× increased capacitance. Here, also, the reshaped ramp of Figure 7.10 will give better linearity.

7.5.1   Spacing change effects

Plate geometries where the pickup is surrounded by fixed, close-spaced stator are more resistant to spacing changes. With the single-stator geometry of Figure 7.6 or Figure 7.8, the performance as spacing varies can be seen by looking at Figure 7.7. The stray capacity, *Cstray*, is constant with spacing while $C1$ and $C2$ change as 1/spacing. For a 10% spacing change and a *Cstray* which is the same as the total pickup capacitance Ct, the apparent position will change by about 10%. This can be fixed in several ways.

In the two-stator-plate geometry of Figure 7.22, the stray capacitance to ground *Cstray* will be minimum. Also, the stator-to-pickup capacitance will be relatively constant as the pickup plate is displaced up and down, especially for small displacements about the centerline. Larger displacements with a low-Z amplifier cause an increase in total capacitance Ct ($Ct = C1 + C2$) which interacts with the pickup amplifier's input capacitance. This graph shows the capacitance Ct vs. z displacement of the two-stator-plate geometry (Figure 7.23).

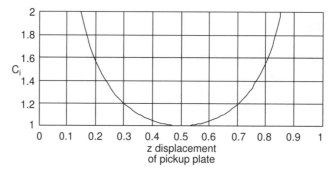

Figure 7.23   z displacement, two stator plates

If the spacing is held within 20% of the center, the variation of total capacitance is held to less than 5% (Figure 7.24).

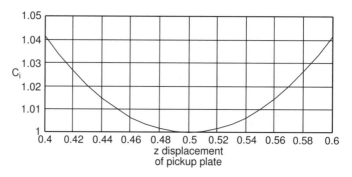

Figure 7.24   z displacement, two stator plates, near center

With large displacements, say, half of the total spacing, as the pickup comes close to either stator plate the total capacitance is dominated by the spacing of the closer plate and changes as 1/spacing, but when the pickup is close to centered, the nonlinear effects cancel over a fairly broad range and vertical axis tilt is first-order compensated. For this to work well, the sum of stray capacitance and amplifier input capacitance should be much less than Ct.

This geometry first-order compensates for tilt in all axes. If the pickup underlaps the stator, translation is also first-order compensated in all axes except the desired x axis.

Guard the pickup

If the pickup is shielded from "looking" at any ground potential (or any harmful noise potential) by a conductive guard plate (Figure 7.25) which is bootstrapped to the pickup's potential, $Cstray$ will be effectively nulled, as seen in Chapters 3 and 4.

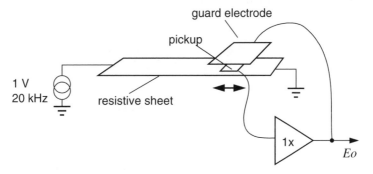

Figure 7.25   Guard electrode

The guard in this example travels with the pickup plate; it is typically larger than the pickup plate, and it shields the pickup plate from stray electric fields and capacitive coupling. If the pickup plate can look in any direction and see either the stator plate or guard, stray capacitance will be totally nulled, provided a zero input capacitance amplifier is used. If the connection between pickup and amplifier is shielded, the shield is connected to the guard so the shield's capacitance does not degrade waveshape. This is a very effective way to null out stray capacitance without adjustment, and with no bad effects on waveshape (Figure 7.26).

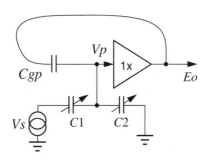

Figure 7.26   Guard electrode equivalent circuit

Cgp is the capacitance between the guard electrode and the pickup plate. The capacity of *Cgp* is larger than the stray capacitance it replaces, but since the voltage across it is forced to be zero by the 1× amplifier, *Cgp* conducts no current. The stray capacity is effectively removed. As the 1× amplifier is typically a low output impedance type, the guard electrode presents a stiff barrier against noise also. If the amplifier gain is, say, 0.99 instead of 1, the effect of *Cgp* will not be completely nulled, but it will be reduced by a factor of 100. *Cgp* will reduce amplifier high frequency gain and rise time.

Neutralizing capacitor

A technique common for vacuum tubes but not often used in semiconductor circuits is neutralization (Figure 7.27). Neutralization compensates for an unwanted input-to-output capacitance by feeding back an out-of-phase output signal to the input through a tunable capacitor, equal in value to the unwanted capacitor. An op-amp implementation is illustrated, as vacuum tubes are becoming difficult to obtain.

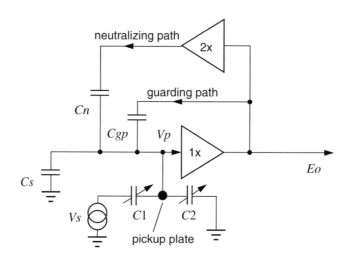

Figure 7.27   Neutralizing amplifier

Here, the bulk of the stray capacitance is compensated by the 1× amplifier driving the guard plate. The remaining stray capacitance (including amplifier input capacitance) is labeled *Cs*. A 2× amplifier provides positive feedback through *Cn*, the neutralizing capacitor. If *Cn* = *Cs*, the effects of *Cs* are removed, and the output voltage will follow exactly the position voltage at the center tap of *C1* and *C2*. If *Cn* is too large, however, the output voltage will be larger than desired and become unstable, so this technique should be used with care and as a last resort if guarding fails to provide adequate performance. Guarding, as is seen in Chapter 10, cannot compensate for some types of internal operational amplifier parasitic capacitances and the neutralizing circuit may be unavoidable.

A single amplifier with a gain of slightly more than one could be used for both guarding and neutralization by allowing *Cgp* to serve both purposes, but the virtue of a 1× amplifier is that its gain can be made very accurate, while a higher gain will depend on resistor tolerance and will necessitate tuning the gain for good performance. The single-amplifier circuit is shown in Figure 7.28.

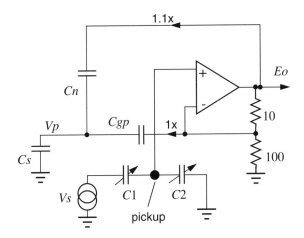

Figure 7.28   Neutralizing with one amplifier

In this circuit, the amplifier produces an accurate 1× output which is used for guarding, and a less accurate 1.1× output which is used for neutralizing. The value of 1.1× may be somewhat arbitrarily chosen between, say, 1.05 and 2. For a gain of 1.1×, Cn's value must be adjusted to $10Cs$, or optionally Cn can be fixed at about $10Cs$ and the amplifier gain tuned for proper neutralization. One minor drawback is that the guard voltage is not as "stiff" as before, with an impedance of 10 Ω, but this still should be acceptable. The amplifier, of course, must have adequate drive capability to handle the 110 Ω load.

7.5.2   Wireless coupling

Capacitive coupling to the pickup plate can be used to eliminate lead wires. The plate geometry discussed above needed three connections, two to the stationary stator plates and one (or two if a guard is used) to the moving pickup plate. All of these geometries can be reversed, that is, consider the pickup stationary and move the "stator" plates instead. This has an advantage, as the pickup is the most sensitive circuit element and shielded cable would be needed to connect to a moving pickup plate, while ordinary stranded wire is fine for connecting to moving stator plates.

Another option can eliminate the wire connection completely. With only slight changes in operation, the connection can be made by capacitive coupling rather than wires. The ramp plate geometry is redrawn in Figure 7.29 with capacitive coupling to the pickup instead of wire pickup.

The pickup now is capacitively coupled to the 1× amplifier through the capacitance between plate C and the extended lower section of the sliding pickup plate. This adds a small (about 1–10 pF) capacitance in series with the 1× amplifier. but if that amplifier has suitably low input capacitance, performance will not suffer.

If the amplifier input capacitance is comparable in value to the pickup capacitance, the amplitude will be cut in half, and an amplitude-independent signal design (see "90°

drive circuits" in Section 10.7.2) is needed. An alternate scheme is to use a feedback amplifier connection (see "Feedback amplifier" in Section 10.3.5).

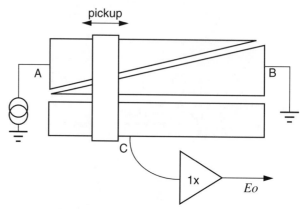

Figure 7.29   Capacitive pickup coupling

Another option is to reverse the roles: fix the pickup plate and couple the drive signal capacitively to the (moving) A and B plates. The moving shield method shown in Chapter 2 may also be wireless, as the shield is normally at ground potential.

8

Multiple plate systems

Capacitive sensors can be quite accurate, with accuracy determined by the precision of the mechanical system. But often some component cannot be precise. If, for example, the capacitive plates are mounted on a mechanical system with poor quality bearings, or the bearings are located at some distance from the transducer, accuracy may suffer as a result of uncompensated tilt or unguarded stray capacitance. When a simple analog approach is not accurate enough, a multiple plate design can be used. This technique applies equally well to linear and rotary arrays, and basically replicates a single-plate analog sensor many times with an added circuit to count plates.

8.1   CAPACITIVE MULTIPLE PLATE SYSTEMS

Multiplate sensors are of two types. Often, the simple three-plate pickup described in Chapter 5 does not have enough capacitance to produce a usable signal-to-noise ratio, or to overcome parasitic capacitance by enough margin for good stability. To remedy this, replicating the three-plate pattern many times and driving each replication identically will increase capacitance, as in the accelerometer described in Chapter 15. This produces a sensor with good resolution, to 5×10^{-12} mm, but poor range of motion, 0.1–1 mm. When the goal is good resolution combined with a long measurement range, often exceeding a meter, the best solution is to replicate a simple plate pattern many times and arrange the drive and pickup circuit so analog interpolation handles the fine position between plates and a digital plate counter adds coarse position information. Multiplate systems include Kosel's linear transducer described in Section 9.2, with a 4×10^{-9} mm resolution and a 3 cm range, and Zhu et al. from Section 9.3 with 1×10^{-6} mm resolution and a range of over a meter. The vernier caliper described in Chapter 18 has a resolution of 2×10^{-4} mm and a 6" range in a compact and inexpensive instrument.

8.1.1   Rotary encoders

Plate shapes

With either capacitive or optical rotary sensors, the normal technique for multiturn high resolution encoding is to build plate patterns (Figure 8.1) which generate sine and cosine waveforms, or triangular waveforms which are used similarly. This allows a rotary encoder to cover more than 360° without ambiguity, and allows a linear encoder to use multiple plates for greater accuracy.

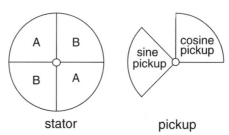

Figure 8.1   Sine/cosine rotary plate pattern

The pickup plates are positioned on top of the stator with small air spacing. This figure shows four stators at 90° and two pickups, but it can be extended to, say, 100 stators at 3.6° to increase resolution, and 50 pickup plates to increase area and thereby decrease noise. The multiple stator plates are connected directly, and two pickup amplifiers and two drive circuits are needed. Again, guard or shield around the pickup plates (not shown) reduces the effect of stray capacity.

With simple pie-shaped plates as shown above, the shape of the demodulated signals produced at a constant mechanical rotation is triangular. A true sine and cosine shape can be obtained by shaping the stator plate as shown in Figure 8.2 so that rotation produces sine waves instead of triangular waves. The shape is defined by setting the convolution of pickup plate and stator plate to the sine of the mechanical angle. If this is done, the mechanical angle is easily calculated as

$$\theta \ = \ \tan^{-1}\left(\frac{\sin \text{pickup}}{\cos \text{pickup}}\right)$$

where sin pickup is the demodulated sine plate and cos pickup is the demodulated cosine signal.

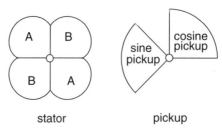

Figure 8.2   True sine and cosine shaping

Angle demodulation

With the pie-shaped plate pattern shown (Figure 8.1), two triangular outputs are produced, called tsin and tcos (Figure 8.3).

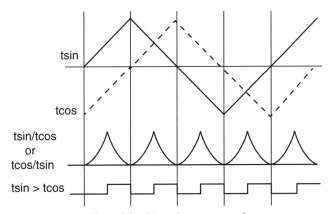

Figure 8.3   Triangular output waveform

tsin and tcos are the demodulated plate signals. The other waveforms are steps in the calculation of angle, as described below. A polar plot of tsin and tcos output signals vs. mechanical deflection angle with the sin and cos plots superimposed is shown in Figure 8.4.

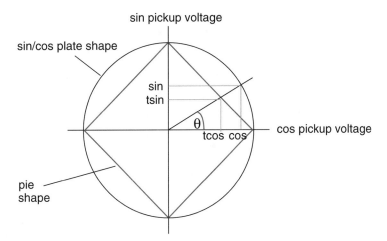

Figure 8.4   Triangular geometry

As the plates are rotated through mechanical angle θ, the true sin/cos (Figure 8.2) plate shape will produce a circular trajectory and the pie-shaped plates (Figure 8.1) will describe a diamond trajectory.

With true sine and cosine waveforms, conversion to angle is done with the inverse tangent calculation. For ease of computation, as the tangent function is infinite at some angles, its reciprocal is used when its value is above 1 to keep the number in bounds. A similar reasoning applies to the triangular equivalent, where the waveform labeled tsin/ tcos or tcos/tsin (Figure 8.3) along with the result of the equality comparison tsin > tcos is used to convert to angle θ, using

$$\theta \ = \ \arctan\left(\frac{tsin}{tcos}\right) \qquad \text{if tsin} < \text{tcos}$$

$$\text{or}$$

$$\theta \ = \ \pi / 4 - \arctan\left(\frac{tcos}{tsin}\right) \qquad \text{if tsin} > \text{tcos}$$

This method requires a little more computation than simply choosing either tsin or tcos based on which is in its linear range, but it is preferred as it is insensitive to absolute changes in amplitude if tsin and tcos are well matched in relative amplitude.

The arctan calculation may be done with a lookup table which also can compensate for any lingering nonlinearities like rounded corners on the triangular waves caused by excessive plate spacing. With two lookup tables, the sensitivity to mismatched relative amplitudes can be compensated.

Linear encoders

Linear multiplate encoders are identical to the rotary encoders described above, except with a linear pattern instead of rotary (Figure 8.5).

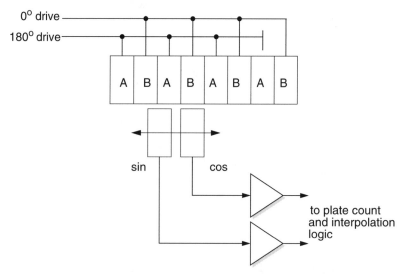

Figure 8.5   Linear multiplate encoder pattern

The electrode pattern above will produce a pair of triangular waveforms, displaced 90 mechanical degrees. These patterns, too, can be shaped to produce an accurate sine/cosine output waveform.

8.2   MULTIPLE PLATE COUNTING

For multiple plate design, after a plate geometry is chosen for either linear or rotary sin-cos signal generation, it is replicated many times in a linear or rotary array. A circuit which counts plates in an up-down counter establishes coarse position, and the arc tan calculation above adds fine position if needed. The plate counting logic is identical to circuits used for optical incremental stripe encoders. These encoders, called also quadrature encoders, have two optical stations mechanically spaced 90° apart, reading out a stripe wheel or grid. This produces two outputs, A and B, shown for a linear sensor (Figure 8.6).

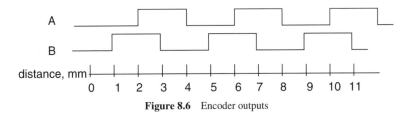

Figure 8.6   Encoder outputs

An up-down counter is used which would store the coarse position, representing 1 mm increments for the case above.

For more precise encoders, the final output position is the sum of the coarse plate count from the up-down counter and a fine analog measurement of the distance to the last plate. This technique is sometimes used for optical encoders, but light source aging, dust, and photodevice tolerances makes analog interpolation inaccurate. It works well for multiple plate capacitive sensors, however, as the analog outputs are quite stable.

A typical multiple plate capacitive sensor outputs 0° and 90° sinusoidal sin-cos or triangular tsin-tcos waveforms, generated by repeated rotary or linear patterns as in Figure 8.1. These waveforms feed both coarse and fine measurement systems (Figure 8.7).

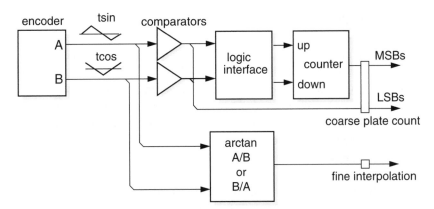

Figure 8.7   Encoder logic interface

The three outputs are added for the complete position determination, with the most significant bits (MSBs) added to the least significant bits (LSBs) to form the plate count, and the arctan calculation added to provide the interpolation fraction.

The logic interface between the quadrature signals A and B and the up-down counter must be carefully designed to avoid miscounts which can happen if noisy edges are present. See, for example, Kuzdrall [1992, pp. 81–87] or Berger [1982, p.128]. The quadrature signal pair at the comparator outputs can be considered as a 2 bit encoder output from an encoder with 90° resolution. Then its state diagram, which shows the relationship between the comparator outputs and the input mechanical angle θ, can be shown as in Figure 8.8.

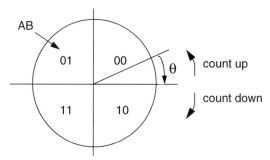

Figure 8.8   Quadrature polar plot

As the input mechanical angle varies from 0 to 360°, the output code successively becomes 00, 01, 11, 10 and repeats.

In an ideal world, the up-down counter could be incremented by the 1 to 0 transition of the A signal if $B = 0$, and decremented by the 0 to 1 transition of the A signal if $B = 0$. Then the A and B signals would be converted from Gray code to binary code and appended to the up-down counter's output as the least significant two bits.

8.3   DEBOUNCE

The debounce algorithm above runs into problems in the real world. With arbitrary mechanical input angles, the A signal may dither at some high frequency and cause continuous activity of the counter, causing excessive consumption in battery powered circuits or synchronization problems for microcomputers. Another more serious hazard: if the arctan calculation produces a fine angle measurement of 359° just as an edge noise glitch on the A signal increments the counter, a 360°, or one plate, error results. These errors can be eliminated by either:

1.   Schmitt triggers on A and B signals instead of comparators
2.   Logic interface design which produces a digital Schmitt trigger effect

Approach 1 will cause the coarse plate counts to lag the fine measurement by an amount proportional to the trigger threshold. Approach 2 is simpler and safer; a mechanical analog illustrates its operation (Figure 8.9).

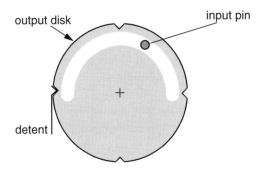

Figure 8.9   Mechanical encoder debounce

The input pin can assume any angle and may dither. The output disk is detented at 90° positions, and allows 180° of input dither before moving to a new position so that only 0°, 90°, 180°, and 270° orientations are asssumed. This analog can be represented by state table (see Figure 8.10), or logic.

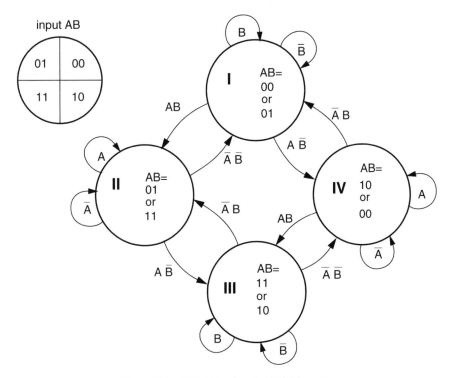

Figure 8.10   State table of mechanical debounce

The "input *AB*" figure shows that the input code is 00 in the first quadrant, 0° to 90°, 01 in the second quadrant, etc. The four output states **I** – **IV** respond to changes in the *AB* input as shown. This state machine can be implemented in logic (Figure 8.11) with the outputs active low. Debounce can also be implemented in logic (Figure 8.11).

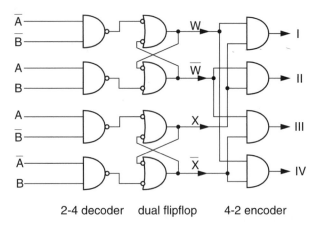

Figure 8.11   Logic diagram of encoder debounce

The noisy *A* and *B* signals produce the clean, debounced gray-coded versions, *W* and *X*, which are recoded into the four debounced quadrant states, **I – IV**. This approach sacrifices 90° of resolution if the fine plate measurement is not used, when compared to an unde-bounced approach. When the fine plate measurement is used, the output is unambiguous and resolution is not lost. The fine plate measurement provides the fraction of 360° mea-surement, while the coarse plate count is derived from the debounced signals.

8.4   TRACKING CIRCUITS

Another way to keep track of position in a multiplate system is by use of tracking logic. A number of tracking algorithms are possible. Here, in Figure 8.12, one pickup and amplifier are used instead of two, and three different sets of plates are driven.

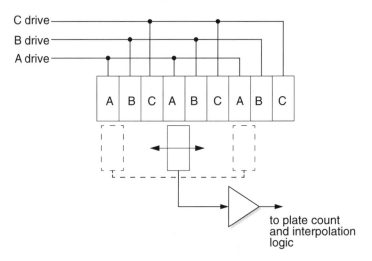

Figure 8.12   An example of tracking logic

8.4.1   Tracking algorithms

A simple tracking algorithm first does an acquisition sweep to see which driven plates are opposite the pickup. In the figure above, plates A and B would be chosen. Then the ratio and the sign of the voltage on A and B are varied until the pickup output is zero. In the figure, that would happen with about four times the voltage on B as on A, and with opposite signs. This ratio is adjusted as needed to null the pickup output to track plate motion; if the plate moves to the right, the voltage on A increases while B decreases until the plate crossed over B, then A drops to zero and C's voltage increases. Generating the tracking signal involves changing the voltage linearly on the three drivers.

8.4.2   Three-phase drive

A second option, properly called a three-phase drive rather than a tracking algorithm, would be to drive the plates with $0°$, $120°$, and $240°$ sine waves or square waves and detect position as the phase angle of the fundamental at the pickup output. No acquisition sweep is needed.

The tracking system has an advantage in that only one pickup amplifier is needed, which saves power in portable applications. Multiple pickup plates as shown in dotted line can be used in either the sin-cos or the tracking system for more output. The disadvantages of tracking systems are the slightly greater complexity of the drive circuit and the extra computation which may limit maximum tracking speed.

8.5   PLATE CONSTRUCTION

Capacitive plates are normally fabricated as a conducting pattern on a dielectric substrate. For additional mechanical or thermal stability, the dielectric is bonded to a more stable material such as stainless steel or Invar. In any case, the critical parameters of plate construction are accuracy, stability, and resistance to environmental factors such as temperature, humidity, and corrosion.

8.5.1   Printed circuit board techniques

The simplest and most convenient way to construct the capacitor plates is often by use of standard printed circuit (PC) technology. The final plate accuracy using this approach will depend on the processes used to fabricate the board as listed below.

Photoplotting

The standard process exposes sensitized polyester film with either a raster-scanned laser or an x-y [Gerber] plotter. Raster-scanned plotters digitize the entire plot and scan the film one line at a time with a small light spot. The older and slower (but no less accurate) vector plotters move a larger, shaped light aperture in vector coordinates. Either method is normally calibrated to an accuracy of 0.0002–0.0008 in over the 22–25 in plotting area, with a minimum pixel size of 0.0001–0.0005 in.

Flatbed plotters and drum types are available. Flatbed plotters are used for glass substrates, but the maximum speed is 1–2 Mpixels/s as compared to the drum plotter's 10–100 Mpixels/s speed.

The speed of the laser plotter allows production of 1× artwork with multiple step-and-repeat patterns directly from the plotter instead of using an intermediate photographic step, thereby improving accuracy.

Phototool

The phototool is the imaged film used for production, and in the case of the laser plotter, it is usually the plotter output directly.

The plotter output is sometimes transferred to Diazo film by contact printing. Diazo film is used as it is convenient and rugged, but exposing Diazo adds an extra photographic step and slightly compromises accuracy. Directly imaged polyester film is the most commonly used medium; 4 mil and 7 mil stock is available with 4 mil stock the most popular, but 7 mil stock is slightly more dimensionally stable.

Polyester is a reasonably stable base, with low humidity-induced creep of 9–14 ppm/°C, a reasonable temperature coefficient of expansion of 15 ppm/°C and excellent mechanical strength. Storage as well as exposure of polyester media should be in controlled temperature and RH conditions; the temperature profile between storage and imaging is important, as polyester has a hysteresis curve for rising and falling humidity. Total imaging tolerances can be about 1 mil in 24 in.

Glass is less sensitive to temperature (about 5 ppm/°C) than polyester and is insensitive to humidity. Glass is used for very accurate substrates, but it is not normally available from a PC house. Some specialized optical shops can generate glass-substrate images. Diamond is still better (1–4 ppm/°C), but its cost renders it unattractive for most nonjewelry applications.

Exposure

The exposure step attempts to exactly replicate the phototool image as a chemical change in the resist coating of the substrate.

The polyester phototool should be exposed to the photosensitive layer of the PC board at the same temperature and humidity as the original exposure. If not, an error of 50–100 ppm is induced by a 10°C temperature change or a 10% change in relative humidity.

Quality PC houses use carefully collimated light to do the exposure, so the accuracy of alignment of the image to the surface of the board is not particularly critical. A vacuum is used to guarantee close contact. Projection exposure is also used, and needs no vacuum system as close contact is not important. The phototool is imaged at 1× magnification with a low f-number reflective lensing system.

The photosensitive layer of the PC board, the photoresist, must first accept an optical image as a chemical change, then, after development with a solvent wash which removes unexposed areas for a negative-working resist, it becomes an etch blocking layer. Liquid and dry film resist types are available. Liquid resists are capable of good performance, but are more difficult to use repeatably; most PC board houses use dry film resist.

Etching

The etching step flows liquid or vapor etchant over the board which etches away the copper in areas where the resist has been removed. Etching requires close control for good accuracy, with the amount of undercut carefully controlled so the image neither grows nor shrinks. Fineline PC board techniques have been developed which allow high yield production of trace widths down to 0.003 in.

PC board substrate accuracy

Once the PC board has been exposed, developed, and etched, some moderate change in substrate dimension may result from chemical absorption. Absorbed solvents can be baked out in drying ovens.

8.5.2   Other techniques

Printed circuit board construction is easy and inexpensive, but other approaches may be needed for unusual applications requiring, for example,

- extreme stability
- extreme tolerance to relative humidity changes
- transparency
- nonplanar shapes
- large size
- very small size

Thin films

Vacuum deposition of a variety of conductive materials to thicknesses between 0.05 and 25 μm can be used to produce capacitive sensor plates. Accurate plate patterns can be defined by sputtering a metal or a compound through a mask onto a suitable substrate, or by coating the insulating substrate with a resist as in PC board fabrication, exposing through a phototool, and developing with solvent wash and etching. The mask is more convenient but less accurate; submicron accuracy is available with resist and phototools using technology developed for integrated circuit fabrication.

Transparent thin films are used in capacitive touch screens for touch-sensitive input application over computer monitors. Many normally opaque metals become transparent in very thin layers; tin oxide and indium-tin oxide are transparent conductors in thicker films and are easier to deposit with good linearity.

Thick films

Thick films are used in producing hybrid circuits and microwave devices. Thick film devices such as conductors and resistors are screen printed onto a smooth substrate, usually alumina ceramic, and fired at 850–1000°C. Film thicknesses are 25–250 μm. The 5–10 mil line definition is not as good as photoetched processes, and no transparent conductors are available, but the process is simple and reasonably inexpensive.

Conductive films

Table 8.1 [Burger, 1972, p. 450] lists conductive films produced by both thin (above the line) and thick film methods.

Table 8.1 Conductive films

Material	Process	Pattern	Thickness μm	Sheet resistance Ω/□
Aluminum	evaporation	mask, resist	0.5–1	0.03–0.06
Titanium	evaporation	mask, resist	0.5	0.1
Silver-chromium	evaporation	reverse metal	0.5	0.1
Gold-chromium	evaporation	reverse metal	0.5	0.1
Copper-chromium	evaporation	mask, resist	0.8	0.02
Platinum-gold-cermet	screen	screen	25	0.08
Silver-cermet	screen	screen	25	0.005
Palladium-gold-cermet	screen	screen	25	0.04

The notation of sheet resistance in $\Omega / \square$, ohms per square, reminds us that sheet resistance is the resistance measured across two opposite faces of a square test pattern and is independent of the dimension of the square.

Resistive thin films

Resistive thin films with capacitive pickup have some applications. See, for example, "Graphic input tablet" in Chapter 19. Producing an accurate high value resistance by vacuum evaporation or thin film sputtering of a conductor is difficult, as the resistivity is a nonlinear function of film thickness with the curve becoming very steep as molecular thicknesses are approached. Most metals have bulk resistivity in the 1–10 μΩ-cm range and would need a very unstable molecular-dimension film thickness to produce an easy-to-drive 1000 Ω resistance. Higher resistivities are available with nickel/chromium alloys or tantalum nitride at 100 Ω-cm [Harper and Sampson, p. 8.28]; these materials produce sheet resistivity of 100–300 $\Omega / \square$. Still higher resistivities can be provided by cermet, a metal/metal oxide composition. Cermets can be thin-film vacuum deposited or screened and fired for thick films, and are most often used for sheet resistivity in the range of 1000–3000 $\Omega / \square$ [Harper and Sampson, p. 8.44].

Resistive thick films

Resistive thick films are often useful in capacitive sensors. The technology of producing resistive thick films is mature and is used in manufacturing very precise potentiometers as well as fixed resistors for application to many different substrates. Many different formulations can be chosen to optimize a particular application.

Potentiometers which use resistive film have an intrinsic sliding noise component due the small contact area of the metal brush and the microscopic surface roughness of the resistive material. This noise component and the added friction and wear of the brush are avoided if a capacitive readout replaces the brush.

A partial list of resistive materials [Burger, 1972, p. 452] appears in Table 8.2.

Table 8.2 Resistive films

Material	Deposition process	Thickness μm	Masking	Sheet resistance $\Omega/\square$	Temp. coeff. ppm/°C
Nichrome	evaporation	0.02	reverse metal, KPR	20–800	+/–100
Tantalum	sputtering	0.05–0.5	reverse metal	10K–20K	+50 to –300
Tin oxide	pyrolysis	0.001–0.5	reverse metal	10–5K	–1500 to +250
Palladium-silver-cermet	screen and fire	25	screen print	1–20K	<200
Rhenium	evaporation	~1000	stencil mask	10–4K	+/–500

Evaporation heats the material in a vacuum chamber until evaporation launches particles which are deposited on a target. Sputtering uses a plasma beam, often argon ions, to blast molecules from a target in a vacuum. Coatings produced in a vacuum chamber are called thin film irrespective of film thickness, and coatings produced by screen printing are called thick film.

Tin oxide is very popular as it is not only a semiconductor, but it is transparent. It is often doped with indium to produce indium tin oxide, ITO, with better physical characteristics than tin oxide. Other deposition methods may be used which do not require vacuum, such as pyrolysis, which exposes a target heated to 300–360°C to gas produced by bubbling nitrogen through indium tin chloride liquid. The gas reacts with oxygen from the air, catalyzed by the hot surface, to form a 1–10 μm layer of indium tin oxide. The accuracy of the resistance can reach 5% with careful process control.

Thick film deposition is discussed by IHSM, the International Society of Hybrid Manufacturers, and thin film by the American Vacuum Society and the Electrochemical Society. Many independent coating laboratories such as Optical Coating Labs in California can handle contract deposition of resistive films.

Metal-filled paint

Several methods are used to coat plastic molded parts with a conductive surface to attenuate RF radiation. One is painting or screen printing the surface with a metal- or carbon-filled polymer with air-drying solvent. Silver is most often used. It is difficult to produce accurate high value resistances with metal-filled paint, and the conductivity is often not adequate for good RF shielding. This method is not useful for very accurate capacitor plate definition, as the limits of screen printing tolerance are on the order of 0.2 mm, but it is convenient and easy to set up. A manufacturer of electrically conductive paint which can cure at low (50°C) temperatures and bonds well to tin oxide or indium tin oxide surfaces as well as to Kapton and glass is Creative Materials Inc., 141 Middlesex Rd., Tyngsboro, MA 01879.

Plastic plating

A variety of metals, including chrome, can be plated onto the surface of plastic parts with good (10–20 lb/in) peel strength. Plating followed by photoimage and etch produces accurate, complex-geometry capacitive electrodes. One company that specializes in molded, selectively plated plastics for electronic interconnection is Mitsui-Pathtek, Rochester, NY, (716)272-3126.

8.6   DIELECTRIC SUBSTRATE

Normal FR-4 glass-epoxy PC substrate works reasonably well as a dielectric for capacitive sensor electrodes. It has the characteristics shown in Table 8.3.

Table 8.3 1/16 in FR-4 PC board characteristics[a]

Dielectric constant @60 Hz	5.0
Dielectric constant @1 MHz	4.6
Dielectric strength, V/mil	360
Volume resistivity, Ω-cm	3.8×10^{15}
Surface resistivity, min., $M\Omega / \square$[b]	10^4
Specific gravity	1.8
Water absorption, %/24 h	0.2
Heat softening temperature, °C	105–125
Tensile strength, lb/in^2	30,000
Coefficient of thermal expansion, ppm/ °C	15–18
Thickness tolerance, mil	+/– 7.5[c]
Maximum acceptable bow and twist, %	0.6 – 2.0[d]

[a] [C. Harper and R. Sampson, *Electronics Materials Processes Handbook,* © 1970, the McGraw-Hill Companies. Reprinted with permission of the publisher]

[b] NEMA minimum; typical values are much higher

[c] For grade A. IPC-L-115 calls out grade B at +/–5 and grade C at +/– 2 mils

[d] Two-sided level C material is 0.8%; level A is 2.0% as measured by IPC-TM-650

Improved performance PC laminates are available from suppliers such as Rogers Corporation, 100 S. Roosevelt Ave., Chandler, AZ 85226, (602)961-1382. Laminates such as Rogers' RO3000 feature very low dissipation factor, <0.0013 @ 10 GHz, 10^{12} Ω/cm volume resistivity, 10^{13} $\Omega / \square$ surface resistivity, and a dielectric constant of 3.0 ± 0.04 with a temperature-induced change of less than 0.5% between –100°C and +250°C.

PC boards can be specified with 1/2, 1 or 2 oz/sq ft copper foil, and 0.1 oz is available at extra cost from specialty suppliers. The thinner foil etches more accurately so undercut during etch is minimized and it is recommended for sensor applications. The thickness of 1 oz foil is 0.00137 in.

In some applications needing excellent stability or extended temperature range operation, PC laminates are available from sources such as Rogers Corporation with superior temperature and physical properties (Table 8.4).

Table 8.4 Properties of some PC board plastics[a]

	E-glass (FR-4)	S-glass	Quartz	Aramid	PTFE[b]
Specific gravity, g/cm^3	2.54	2.49	2.20	1.40	2.2
Tensile strength, kg/cm	350	475	200	400	210
Young's modulus, kg/cm^2	7400	8600	7450	13,000	3500
Maximum elongation, %	4.8	5.5	5.0	4.5	150
Specific heat, cal/g°C	0.197	0.175	0.230	0.260	
Thermal conductivity, W/m–°C	0.89	0.9	1.1	0.5	
Coef. of temp. expansion, ppm/ °C	5.0	2.8	0.54	–5.0[c]	99
Softening point, °C	840	975	1420	300	
Dielectric constant at 1 MHz	5.8	4.52	3.5	4.0	2.0
Dissipation factor at 1 MHz	0.0011	0.0026	0.0002	0.001	<0.0002

[a] [Senese, 1990]

[b] [Data from C. Harper and R. Sampson, *Electronic Materials Processes Handbook*, © 1970, the McGraw-Hill Companies. Used with permission of the publisher]

[c] Along fiber axis

E-glass

This is the normal G-10 or FR-4 substrate used for 95% of PC boards. It is made from woven continuous-fiber borosilicate glass in an epoxy binder. Specifications are listed in IPC-EG-140, from the Institute for Interconnecting and Packaging Electronic Circuits.

S-glass

This is a similar glass except with a considerably better performance at high temperatures. It is more expensive. Specifications are in IPC-SG-141.

Quartz

Quartz (fused silica) glass is more expensive, more stable, and more abrasive, making PC drilling more difficult. Specifications are in IPC-QF-143.

Aramid

Aramid fiber, also known under DuPont's trade name of Kevlar, has a negative temperature coefficient which, when combined with epoxy's positive coefficient, can produce very low total temperature coefficient. It is expensive and difficult to drill in conventional PC boards, although it is used extensively in flexible thin-substrate circuits.

The dielectric constant of PC boards will increase after water absorption. Several different types of thermoset resins of the type used for PC board laminates were conditioned by boiling in water for 24 h, and exhibited increases in dielectric constant of 10–30% [Shimp, 1990]. This reference also tested different resins for dielectric constant stability with temperature, and found the dielectric constant of normal epoxy laminate within 2% from 0–100°C but increasing by 15% at 200°C. Newer resin types such as AroCyB cyanate ester homopolymer have a dielectric constant which is flat to 1.5% over 0–200°C.

1/16 in glass-epoxy PC boards are normally specified to have a curvature of less than about 0.01 cm/cm. Thicker stock is used for designs needing better flatness than this, or the PC material can be bonded to a metal backing.

Table 8.5 Properties of various dielectric substrate materials[a]

Material	Dielectric constant	Volume resistivity Ω-cm	Surface resistivity $\Omega/\square$	Dielectric strength, V/mil
Air	1	--	--	--
ABS	2.9–3.3	10^{15}–10^{17}	--	300–450
Alumina	8–9	10^{14}–10^{15}	--	250–400
FR-4 PC board	4.6	3.8×10^{15}	--	--
Glass (common)	8	10^{15}–10^{17}	10^{12}	--
Glass (fused silica)	3.78	$>10^{19}$	2×10^{14}	1500
Glass-epoxy	5–7	--	--	--
Nylon	6.0	10^{14}–10^{15}	--	300–400
Polyethylene, med. density	2.2–2.3	10^{15}–10^{18}	--	450–1000
Polyimide	3.5	10^{16}–10^{17}	--	400
Polystyrene	2.4–2.7	10^{17}–10^{21}	--	300–3000
PTFE	2.1	$>10^{18}$	--	480
Solder mask, dry film	3.6	$>3 \times 10^{14}$	$>7 \times 10^{13}$	--
Titanium dioxide	14–110	--	--	--
Water	80	--	--	--

[a][Harper and Sampson, 1970, pp. 1.20, 3.4–3.11]; *Reference Data for Engineers*, pp. 4–20

PTFE

Polytetraflouroethylene (Teflon^{TM}) is used for specialized microwave PC boards because of its low dielectric constant, 2.1. It also has a low dissipation factor, 0.8×10^{-4}, excellent volume and surface resistivity, and low friction, and it is one of just a few plastics with zero water absorption. These properties are valuable for capacitive sensors if the additional cost of PTFE stock is acceptable.

Dielectric layers

A dielectric layer is sometimes used in variable capacitors to increase capacity by virtue of its dielectric constant and to allow close spacing without danger of shorting. Also, the dielectric layer acts to stabilize the plate structure against mechanical resonances.

Dielectric layers are also useful for capacitive sensors for the same reasons. The increase in capacity causes a beneficial increase in signal strength. The ideal dielectric would have a large and stable constant, low friction, and good resistance to wear. Unfortunately, the lowest friction material, PTFE, also has the lowest dielectric coefficient and is difficult to manufacture in thin gages, although Teflon/anodize coating is available for aluminum in thin layers. (See Table 8.5.) For more detail on the physical characteristics of glass, see Harper and Sampson [1970, Chapter 3.]

9

Miscellaneous sensors

This chapter presents a sampling of capacitive sensor designs for a variety of miscellaneous transducers. Just about any measurement task can be handled by converting the variable to be measured into mechanical displacement of capacitor electrodes; Jones [1973] points out that an excellent thermometer could be built by capacitively sensing the change in size of a brass block with temperature.

9.1   PRESSURE SENSORS

Capacitive pressure transducers are generally built with spacing-change sensors, governed by the usual equation

$$C = \varepsilon_0 \varepsilon_r \frac{A}{d} \qquad\qquad 9.1$$

When the plates are parallel and displaced by Δd, the change in capacitance ΔC is

$$\frac{\Delta C}{\Delta d} = -\varepsilon_0 \varepsilon_r \frac{A}{d^2} \qquad\qquad 9.2$$

Hence as dimensions are scaled to submillimeter silicon size, the capacitance scales linearly, but the percent change in capacitance with displacement does not change. Capacitive pressure sensors are displacing piezoresistive pressure sensors because of lower power requirements, less temperature dependence, and lower drift.

9.1.1   Differential pressure transducer

A typical differential pressure transducer construction is shown in Figure 9.1.

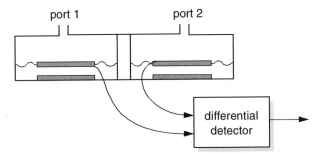

Figure 9.1   Differential pressure transducer

9.1.2   Absolute pressure sensors

A medical implant pressure sensor constructed with discrete components [Puers, 1993] has been developed by a pacemaker company for automatic defibrillation, and is optimized for small size, low power consumption, and high sensitivity (Figure 9.2).

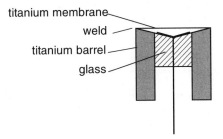

Figure 9.2   Pressure transducers

A titanium barrel is filled with a glass cylinder and sputtered with a gold layer to form the fixed electrode, and a thin titanium diaphragm is welded to the periphery to form the movable electrode. The total capacitance is 1.5 pF.

An alternate construction method [Puers, 1993, p. 96] is to use a flexible silicone rubber dielectric to support the moving plate. This technique has been applied to force transducers as well as pressure transducers, and has the advantage of a sensitivity improvement due to the dielectric constant increase which compensates for its lower compliance compared to air.

9.1.3   Silicon pressure sensors

Silicon based capacitive pressure sensors have these advantages over alternate construction methods:

* Silicon has excellent elastic properties and excellent stability
* Capacitance conversion is a stable and noise-free conversion method well-matched to IC fabrication processes and batch technology
* Capacitive sensors have been fabricated with a gap of 1 μm, for a very acceptable 8.85 pF/mm^2 electrode capacitance

A review of pressure sensors fabricated on silicon shown in Appendix 1 [Puers, 1993, pp. 97–104] covers 25 different devices from many different laboratories, with dimensions down to 0.4 × 0.5 mm and capacitances on the order of 0.3–25 pF. Accelerometers are also listed, with sizes down to 0.3 × 0.1 mm and capacitance down to 0.004 pF. The earlier devices listed use primarily oscillator circuits for demodulation, with a later trend to switched capacitor CMOS with one device using NMOS; synchronous demodulators are beginning to appear in later circuits.

Linearity

Performance of silicon pressure sensors without adding linearity compensation is good. A nonlinearity is caused by the nonparallel deflection of the moving plate, which may form a dome shape when displaced, as the gap is of micron dimensions for sensitivity the nonlinearity can be significant. Adding a guard ring (the Kelvin guard) to the smaller of the two electrodes and reducing its area to 36% of the diaphragm area decreases sensitivity but improves linearity to 1.5% [Artyomovet et al., 1991], with a pressure-induced capacitance change of 2.5×.

Another approach is to stiffen the center section of the moving diaphragm with a boss [Schnatz et al., 1992, p .79] (Figure 9.3).

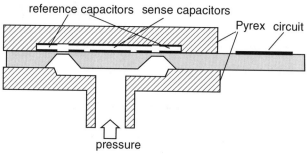

Figure 9.3   Bossed diaphragm

The demodulation circuit for the sensor above places the capacitor in the feedback path of the input op amp to linearize it, as discussed in Chapter 4, adds the central boss as shown to ensure that the plates are parallel, and includes a bandgap temperature compensation for temperature effects, primarily due to the different temperature coefficients of silicon and Pyrex glass. Two reference and two sense capacitors and a differential CMOS switched-capacitance amplifier were used for a very accurate 0.3% nonlinearity over an input pressure range of 0 –30 kPa and temperature dependence of 13% p-p over the range –60 to +140°C.

9.2   3 cm LINEAR TRANSDUCER

A linear transducer with multiple sine-cosine plates on 41 µm (0.0016 in) centers was designed to replace diffraction grating and laser interferometer measurement systems at lower cost [Kosel et al., 1981] for precision positioning uses like semiconductor step-and-repeat imaging. An air bearing support is used for zero friction. The electrode design is similar to the 6 in caliper (Chapter 15) except a factor of 60 smaller, with 0.5 μm etched aluminum electrode fingers on a 6.3 mm glass substrate and a protective overcoat of 1 μm photoresist. The transducer has these characteristics:

- Position uncertainty of 4 nm
- 3 cm effective linear range
- Operating frequency 870 kHz
- Plate separation 15–23 μm
- Transducer capacitance 10 pF

The position uncertainty is due to resistor thermal noise, and was improved by tuning the detector circuits to the operating frequency with a Q of 30.

Kosel presents an analysis of the coupling efficiency of the gaps, with computer simulations which show that the coupling of about 10% in air deteriorates to 3% with glass substrate, but improves to 16% with a dielectric (resist) layer over the electrode-glass layers. A formula is derived for the output voltage of the transducer as a function of electrode finger dimensions, gap size, and linear displacement. With this electrode geometry, showing only one of the two short gratings (Figure 9.4)

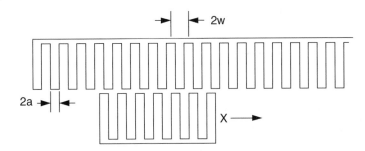

Figure 9.4   3 cm linear transducer [© 1981 Institute of Electrical and Electronics Engineers]

and approximating the finger dimensions with these values

$$w = 20 \ \mu m$$

$$a = 10 \ \mu m$$

$$d = 19 \ \mu m = \text{gap width}$$

the formula for output voltage as a function of linear displacement is

$$V_j = \frac{\ln\left[\frac{\left(\cosh\left(\frac{\pi \cdot d_i}{2 \cdot w}\right)\right)^2 - \left(\sin\left(\frac{\pi \cdot X_j}{2 \cdot w}\right)\right)^2}{\left(\cosh\left(\frac{\pi \cdot d_i}{2 \cdot w}\right)\right)^2 - \cos\left(\frac{\pi \cdot X_j}{2 \cdot w}\right)}\right]}{\ln\left[4 \cdot \left(\frac{w}{\pi \cdot a}\right)^2 \cdot \left(\sinh\left(\frac{\pi \cdot d_i}{w}\right)\right)^2 + \left(\sin\left(\frac{\pi \cdot X_j}{w}\right)\right)^2\right]}$$

And a numerical evaluation for output voltage vs. linear displacement X, for two different gaps of 10 and 20 µm, is shown in Figure 9.5.

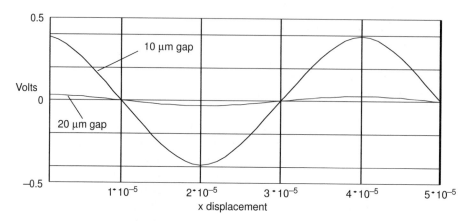

Figure 9.5   3 cm linear transducer output vs. displacement
[© 1981 Institute of Electrical and Electronics Engineers]

The extreme sensitivity of this system to gap width is shown in Figure 9.6.

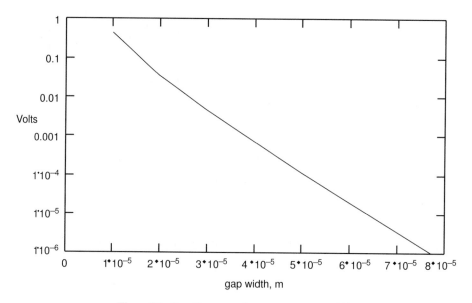

Figure 9.6   3 cm linear transducer output vs. gap width

This graph shows the typical rapid falloff of signal with spacing and the need to keep gap size smaller than electrode size for good performance. The design challenge for this transducer is to keep the gap very well controlled so that the output voltage is not modulated, or include a reference level so that the amplitude modulation due to gap change can be compensated. Increasing electrode size for higher depth of modulation would ease the gap tolerance requirement, and the reduction in sensitivity due to the increased pitch should be compensated by the increased modulation depth.

9.3   LINEAR MULTIPLATE MOTION SENSOR

A prototype sensor was built by Zhu, Spronck, and Heerens [Zhu et al., 1991] with the following specifications:

Coarse plate pitch . . . . 2 mm
Fine plate spacing . . . . 250 μm
Substrate construction. . . . . PC board on steel backing
Electrode capacitance . . . . 0.75 pF
Resolution . . . . 1 μm
Stability . . . . <1 μm
Measurement speed. . . . . 500 mm/s
Linearity . . . . 50 μm

The sensor was designed to be capable of a measurement range of a few meters, to be insensitive to misalignment, dirt, and electromagnetic disturbances, and to be manufacturable at low cost.

A novel technique is used for measuring position. The fine grid electrode configuration with 250 μm pitch opposes a coarse grid of plates with 2 mm pitch with a small air gap of 250 μm (Figure 9.7).

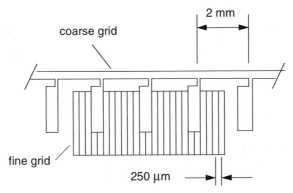

Figure 9.7   Linear multiplate sensor electrodes [Reprinted from *Sensors and Actuators*, A25–27, Zhu et al., "A Simple Capacitance Displacement Sensor," p. 266, 1991, with kind permission from Elsevier Science S.A., Lausanne, Switzerland]

The fine plates are connected together in groups of 8, plate 1 is connected to plates 9 and 17, etc., plate 2 is connected to plates 10 and 18, etc., so that their signals will average out small irregularities and will sum together for more signal amplitude. The fine plates are scanned with a clock at an adequately high frequency for the anticipated fastest translation speed. A block diagram of the electronics shows the scanning method (Figure 9.8).

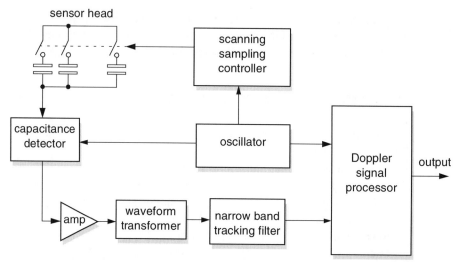

Figure 9.8   Linear multiplate sensor block diagram [Reprinted from *Sensors and Actuators*, A25–27, Zhu et al.,
"A Simple Capacitance Displacement Sensor," p. 268, 1991, with kind permission from
Elsevier Science S.A., Lausanne, Switzerland]

As the electrode air gap is about the same as the pitch of the fine plates, the fine plates will
exhibit a capacitance which is a defocused image of the coarse plates (Figure 9.9).

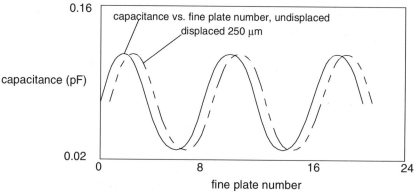

Figure 9.9   Capacitance of fine plates [Reprinted from *Sensors and Actuators*, A25–27, Zhu et al.,
"A Simple Capacitance Displacement Sensor," p. 268, 1991, with kind permission from
Elsevier Science S.A., Lausanne, Switzerland]

The output of the fine plate scanner is an AC signal which varies in phase as shown above,
directly as the relative position of the two sets of plates, so that a displacement of 250 μm
causes the capacitance-vs.-electrode-number curve to move by one electrode. The scanner
continuously scans through the 24 fine electrodes, so the scanner output is an AC signal
which can be compared in phase to the oscillator output to determine mechanical displacement. The small capacitance changes, a fraction of a pF, mean that the usual guarding pre-
cautions need to be exercised, and the preamp needs to be very close to the electrodes.

The frequency shift of the AC signal is proportional to the velocity of the motion, similar to Doppler frequency shift. A variety of techniques are available to convert a frequency shift into position by measuring phase; the graphic tablet of Chapter 19 shows an example.

This sensor illustrates a novel method of position detection. It is a hybrid of two techniques which have been described, the tracking method and the sine-cosine method, and should offer similar performance, with the additional complexity of the sampling circuits compensated by less-critical analog circuits.

9.4   MOTOR COMMUTATOR

DC brushless motors are increasingly used to replace AC motors, steppers, and brush-type motors, as they have superior efficiency, long life, smooth torque delivery, and high speed operation. DC brushless motors reverse the construction of a DC brush motor, with wire coils outside and a permanent magnet rotor inside. The inherent cost of the motor is lower than brush motor cost, but as brushless motors are not built in large quantity as a standard part, the actual cost will be higher for quantities of under 100K/yr. Total system cost may be higher due to the more complex drive electronics needed.

DC brushless motors have performance similar to brush motors, with a slightly higher efficiency as the brush voltage drop is not a factor. Also, the small brush friction component of brush motors is eliminated. Brushless motors use a variety of commutators to do the function of the segmented commutator of the brush motor; this section is a proposal for the use of capacitive transducers for brushless motor commutation. Most of the circuits and systems shown in this book have been built in production or at least prototype form, but this idea has not been tested.

9.4.1   Commutation

Where commutation of DC brush motors is handled by the brushes, commutation for brushless motors is done by external electronic switches controlled by rotor position. Two or more poles can be used, with more complexity but higher torque for the higher pole count. The normal setup uses a three-phase winding and a two- or four-pole magnet, with three Hall effect devices which sense rotor position magnetically and actuate the proper coil winding switches.

9.4.2   Ripple torque

The motor designer can choose between designing the magnetic structure for a linear torque vs. position relationship which may have a large switching transient, or a sinusoidal torque vs. angle with more ripple but less commutation difficulty. For linear torque designs, the torque transient at commutation may be greater, as the normal Hall effect device is inaccurate enough to produce considerable error in commutation angle. This presents no problems at high speed, as inertia keeps speed reasonably constant, but in low speed servos this effect can cause jitter.

Transient torque and ripple are reduced with sinusoidal torque-vs.-angle motors and "soft" commutation which uses linear amplifiers instead of switches, with a more accurate rotor position sensor replacing the Hall effect devices. The best performance uses an absolute position encoder and linear amplifiers, and determines the best commutation angle by powering the windings with DC and measuring the rotor angle.

Capacitive sensors are preferable for high performance applications to Hall effect sensors, as they are considerably more accurate and less expensive, and can be easily integrated onto the PC board normally used to support the winding.

Brushless motors, as with brush motors, can be built either in disk or cylindrical geometry. Disk construction is used for head drives in VCRs and platter drives in computer hard disk drives. These motors are usually built on a printed circuit board substrate, and use Hall effect, back EMF sensing, or magnetic coupling to sense motor speed for commutation. The Hall effect commutators are not particularly accurate, and would not provide smooth, accurate, linear commutation signals for smooth motion, and other magnetic commutators do not work at slow speed.

Brushless motors with disk construction and a capacitive rotor position commutator (Figure 9.10) can be used at slow speed; they will have accurate, linear commutation and this technology can be used for a high performance assemble-it-yourself motor. This configuration is quiet and smooth, and ripple torque is easily compensated with a table lookup in microcomputer systems.

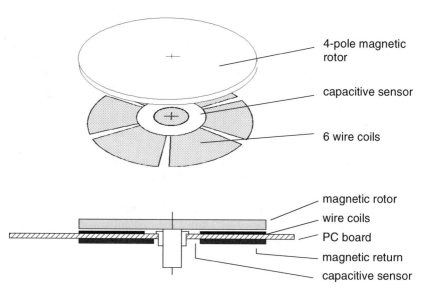

Figure 9.10   Brushless motor with capacitive position sense

The capacitive position sensor does double duty for both motor commutation and shaft position measurement. It uses a conducting pattern on the rotor which may be a metalized paper or plastic label. A typical electrode pattern which can be used directly for linear commutation is shown in Figure 9.11.

This pattern can be 2 cm or less in diameter. Three shaded rectangular pickup plates are shown, each of which can feed a high impedance amplifier and a linear motor drive amplifier. The rotor plate can be connected to capacitively coupled rings to handle drive voltages without need for wire connection. This pattern will generate three-phase sine waves for a brushless motor drive (Figure 9.12). Rounded trapezoidal shapes would be used if the motor magnetic structure were more linear.

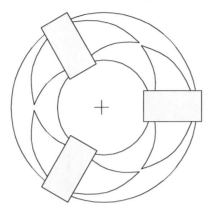

Figure 9.11   Brushless motor position sensor

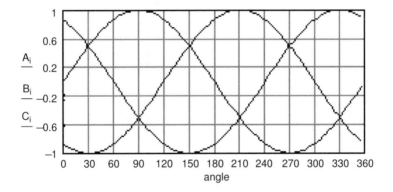

Figure 9.12   Brushless motor drive waveform

These outputs can also be converted to an angle for shaft position measurement as shown in Figure 21.2. One efficiency-increasing possibility is to generate a PWM signal from the sensor clock which is applied to the motor coils instead of a linear voltage.

9.5   WATER/OIL MIXTURE PROBE

Accurate measurement of the percentage of water in oil is extremely important for operators of off-shore oil drilling rigs and for refineries. Capacitive probes have been used for this purpose since 1982, but early units were unstable and temperature dependent. The principle behind these probes is to measure the dielectric constant of oil and of water separately at the operating temperature of the sensor and then to measure the dielectric constant of the mixture. A change in the relative volume of the two components produces a linear change in the measured dielectric constant.

9.5.1   Simple probe

A simple water/oil probe that does not work well is shown in Figure 9.13. This probe will have several problems.

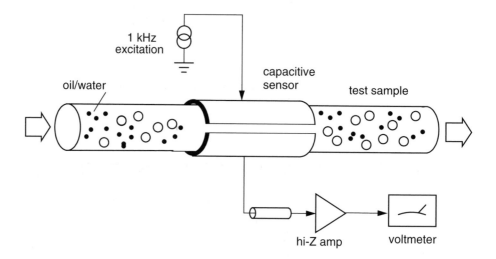

Figure 9.13   Simple oil/water probe

Parasitic capacitance

The probe capacitance will be on the order of 1–10 pF. This will be swamped by capacitance from the back side of the pickup plate and coax capacitance at 50 pF/m. As coax capacitance will have a respectable temperature coefficient, this probe's accuracy may be limited to 30% or so with temperature variation.

Temperature dependence of dielectric constant

The dielectric constant and loss tangent of water is shown in Figure 10.4, and the temperature dependence of these parameters is shown in Figure 10.5. Extrapolating these figures shows that water will be represented as a temperature-dependent resistor at the excitation frequency of 1 kHz, with a high temperature coefficient. This effect, also, will severely limit the performance of the simple probe.

9.5.2   Improved probe

Van der Linden [1988] describes an improved version of an oil-water probe which can measure concentrations of water from 100–1000 ppm with a resolution of 10 ppm (Figure 9.14). The exact capacitance can be calculated as described by Heerens et al. [1986].

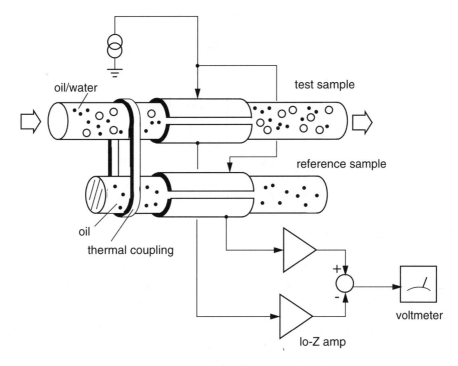

Figure 9.14   Improved oil-water probe

Parasitic capacitance

The guarding principles discussed in Chapters 2 and 3 are used to eliminate the effects of parasitic capacitance. If a low-Z amplifier is used, the coax shield can be grounded without adding coaxial cable capacitance to the measurement and the back side of the pickup electrode can be shielded with ground, so only the cell capacitance will contribute to the signal. The ground shield should be spaced as much as possible, as pickup electrode coupling to this ground decreases signal to noise ratio as seen in Chapter 12.

Temperature dependence

The temperature dependence can be reduced considerably with this two-probe circuit, one measuring the liquid under test and the other measuring a reference liquid. In this case, where small concentrations of water are expected, the reference liquid is 100% oil. Couple the test and reference probes closely for thermal matching (Figure 9.15).

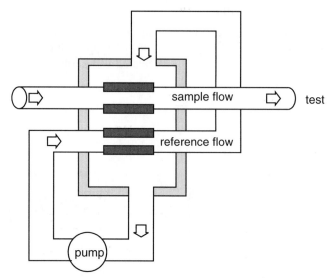

Figure 9.15   Thermal coupling of test and reference cells

9.6   TOUCH SWITCH

A very simple (one resistor) but reliable switch can be built for finger activation if a micro-computer with a comparator input is available (Figure 9.16). Plate *b* can be a 2 cm square copper pattern on the top of a printed circuit board. Plates *a* and *c* are on the bottom side, and with 0.062" glass-epoxy printed circuit board, will couple with about 6 pF to the top plate. A 5 V pulse on plate *a* is coupled through plate *b* to sensor *c*, and can be detected by a microcomputer such as the 80C552, a version of the 8051 made by Philips, with a com-parator input. The received signal will be about a volt; when the top plate is touched by a finger, the coupling drops and the received signal drops to 0.1 V or so. A plastic actuator can replace the human finger for keyswitch applications; the actuator can be nonconducting and increase coupling by its dielectric constant, or decrease coupling if it is conductive.

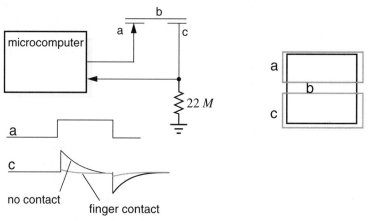

Figure 9.16   Finger-touch switch

9.7   KEYPAD

The techniques of two-plate proximity sensing can be extended to a multiplexed keypad. As the different keys can have different capacitances due to interconnection capacitance variation, the received signal comparator threshold is adapted to the key capacitance using a calibration sweep through the keys during power-up when no key is touched.

Kronos, Inc., of Waltham, MA, used a keypad of this type in a wall-mounted computerized time clock. The keypad was used only by the operator, so it was normally blacked out. The blackout construction used a transparent keypad with a transparent conductor, tin oxide, on 1/8 in thick polycarbonate plastic. Keys and sense electrodes were deposited in this pattern (Figure 9.17). A section view shows the construction (Figure 9.18).

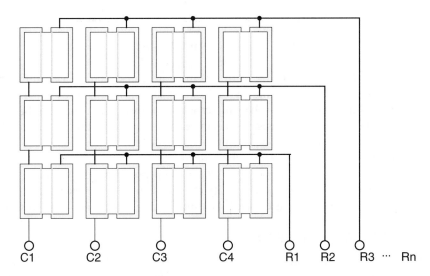

Figure 9.17   Keypad plate pattern

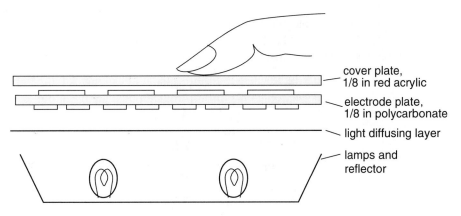

Figure 9.18   Keyboard construction

The column electrodes *C1–C4* and the row electrodes *R1–Rn* are multiplexed using CMOS switches, type CD4051 (Figure 9.19).

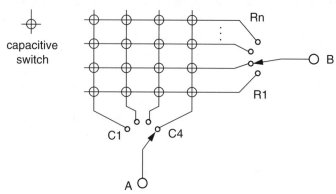

Figure 9.19   Keyboard multiplex

The control microcomputer sequences through each switch, measures its capacity using a technique similar to the single-switch circuit above, and compares it to the threshold value stored for that switch. As the stray capacity for this construction can be greater than the measured capacity, a reasonably accurate digital representation may be needed, perhaps 4 or 5 bits, to discriminate between the considerable variation of stray capacitance between the different switches. Alternately, the signal electrodes can be individually adjusted to add intentional stray capacitance which compensates for the variation. This adjustment will not be needed if signal capacity change is much larger than circuit and stray wiring capacity or guards are used to null out stray wiring capacitance and avoid the adjustment.

9.7.1   CMOS and ESD

Electrostatic discharge (ESD) must be correctly handled for keypads. The human body can be modeled as a capacitor to ground of 350 pF in series with a 1 kΩ resistor (rather a trivial representation of the complex human condition), and in dry climates up to 25 kV of charge on this equivalent capacitor can be discharged through the keypad. This may produce a circuit transient which is interpreted as a logic pulse or in extreme cases the discharge can destroy integrated circuits.

CMOS devices, like the older metal-gate CD4000 series or the newer silicon-gate HC4000 parts, are protected against ESD with this internal circuit (Figure 9.20).

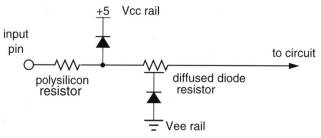

Figure 9.20   ESD protection circuit

The JEDEC-registered maximum voltage rating for CMOS input circuits is $-0.5 < Vin < Vcc + 0.5$, but some manufacturer's specifications are more forgiving, allowing the input voltage to swing as much as 1.5 V over the power rails. The diodes are rated at 20 mA, so ESD using the human body model must be limited to 20 mA continuous. For the short transients of ESD events, CMOS inputs can normally withstand 2.5 kV, still much less than could be expected from an actual discharge. Also, the channels of multiplexers (the analog switch terminals) are more sensitive than normal logic inputs. If the DC voltage on a multiplexer channel is more than a few tens of mV outside the power rails, internal parasitic bipolar transistor structures are turned on, excess power supply current is taken, and other multiplexer channels that are normally off in the same device begin to turn on.

Keyboards can have a protective conducting plate which absorbs an ESD discharge and conducts it to chassis ground, or a suitably thick plastic plate as in the design above prevents the discharge from occurring, as the dielectric strength of most plastics is in the 10–20 kV/mm range. Often, high current diodes such as the 1N4001 are used to shunt ESD discharges to the power rails in order to protect CMOS inputs from damage, but these diodes will not prevent a logic transient. Capacitive keypads have an advantage over other types, as a keypress can be sensed through a thick plastic material which prevents breakdown from occurring, even with a 40 kV charge. Because of the probability of an ESD discharge of producing a logic transient, software which handles capacitive sensor inputs should discriminate against short, less than 100 μs, logic glitches.

9.7.2   Keypad software

A keypad such as the above has some characteristics which may be handled in software.

Ghost keys

In any multiplexed keypad such as Figure 9.19 without sneak path protection, the key contact can conduct in either the normal direction or, when three or more keys are pressed simultaneously in a particular pattern, the contact conducts in the reverse direction, producing unexpected results. If three keys are simultaneously pressed with the keys forming three of the four vertices of a rectangle, the key at the fourth vertex may be read as being pressed also. This "ghost key" can be handled with mechanical keypads by use of a diode on each key to prevent reverse current flow, but on a capacitive keypad this is not an option. One option on a capacitive keypad is to take advantage of the lower capacity through three keys in series.

Ghost key prevention can be handled by software, however. The software must keep track of keypress history, and if three keys such as $R1$-$C1$, $R2$-$C1$, and $R1$-$C2$ are pressed and held, in that order, the software will also receive the ghost key $R2$-$C2$ and be unable to determine whether the third key pressed was $R1$-$C2$ or $R2$-$C2$. To resolve the ambiguity the software must detect that the number of keys pressed has increased from 2 to 4 and look for a rectangular pattern. If such a pattern is detected, the third keypress is not resolved until the first key is released; then the third keypress is known and is transmitted.

ESD transients

The short transient produced by an ESD event is handled in software by resampling each input after a millisecond or two to reject transients. Interrupt-driven software can handle ESD problems as well as switch bounce.

9.8   TOUCHPAD

The mouse serves personal computers well as an input pointing device, but it has not been as popular for portable computers, where a number of solutions have been tried. Recently, small capacitive touch pads integrated into laptop computers are becoming popular. Several manufacturers make these products, including Philips, Symbios/Scriptel, Synaptics, and Alps.

Synaptics' touch pad is smaller than a business card, with a smooth mylar surface. It handles cursor control by finger position, and also substitutes for mechanical mouse switches by interpreting a finger tap as a switch closure. The standard size is 50 × 66 mm, with smaller sizes available, and thickness is less than 5 mm. The resolution is better than 500 dots/in, so that fine cursor control is achieved by rocking the finger slightly. As the device is sensitive to the total finger capacity which will vary with pressure, a degree of z-axis pressure response is available, which could be used to modulate line width in a paint program or control scroll speed in a word processor.

Electrostatic discharge performance is good, with soft failures occurring at 15 kV and hard failures at 20–25 kV.

9.8.1   Sensing Technique

The sensing technique, an analog version of the keypad above, uses an x-y grid of wires etched on the upper two layers of a four-layer printed circuit board (Figure 9.20a).

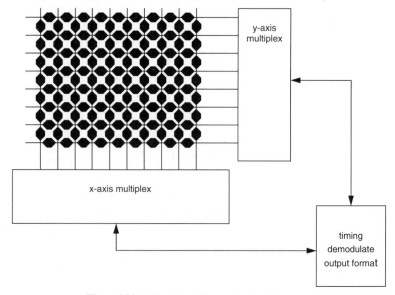

Figure 9.20a   Example of the sensing technique

Capacitive coupling is improved by widening the traces, and finger capacitance is sensed as a decrease in coupling between adjacent electrodes caused by the shielding effect of the relatively high capacitance of the human body.

The electronics are surface-mount components on the bottom side of the board, including a mixed-signal ASIC and one or more digital chips if needed for special interface requirements. A combination of digital sensing and analog interpolation between wires is used to accurately determine finger location. The connections from the ASIC to the electrodes are designed to have equal capacitance despite unequal trace length.

Both the electronic and mechanical complexity of the touchpad are less than trackballs, and with no moving parts service life is much higher.

9.8.2   Accurate two-dimensional sensor

A more accurate two-dimensional transducer can be built with similar techniques if the finger is replaced with a close-spaced plane pickup electrode. Bonse et al. [1994] describe an 85×60 mm unit using a 14×10 array of 6 mm^2 square electrodes on the stator and a moving 3×3 electrode pickup. This device, fabricated with printed circuit board technology, demonstrated a 100 nm resolution and 200 nm repeatability. Sine wave excitation and a synchronous demodulator were used, and the major contribution to inaccuracy was determined to be the stability and tolerance of the printed circuit board.

9.9   SPACING MEASUREMENT

The CMOS 555-type timer and the RC circuit form an oscillator which changes frequency as $1/RC$. As the capacitance C is proportional to 1/spacing, the output frequency is directly proportional to plate spacing and may be easily measured by counting pulses for some fixed interval of time (Figure 9.21).

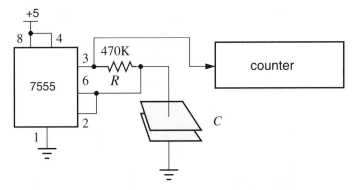

Figure 9.21   Spacing measurement system

If the plates are large and opened, a proximity detector is implemented which detects either dielectric or conductive material in the vicinity of the plates.

This circuit is not too accurate or sensitive, as the input capacitance of the 7555's pins 2 and 7 is about 15 pF and changes slightly with Vcc. For better performance, add a driven guard as described in Figure 4.3.

9.10   LIQUID LEVEL SENSE

Liquid level sensing is easily handled by capacitive sensors. Conductive sense electrodes are applied to the sides of the tank, inside the tank if it is conductive or outside if it is a nonconductor, and the capacitance between electrodes is measured. As the liquid level is measured by its volume characteristics rather than its surface properties, the liquid level can be measured reasonably well in the presence of sloshing which would confuse a standard float sensor, and if the electrodes are large (the complete top and bottom, say) a good zero-g or high-slosh liquid volume sensor is implemented.

The Levelite Store, Port Huron, MI (800)975-3835, specializes in level measurement, offering hundreds of different sensor types using every available technology. Levelite says this about capacitive level sensors:

- Unquestionably [the capacitive level sensor] is the Levelite Store Catalog's most versatile switch, used by industries all over the world involved with chemicals, foods, and wastewaters
- Ideal for dirty liquids and slurries as well as difficult dry solid applications
- Will detect interface between oil and water
- Sensor-Guard™ probe helps ignore buildup on the sensor
- Operates independently of specific gravity, conductivity, or viscosity
- A dielectric constant of 2.0 or more is recommended

The capacitance between electrodes can be measured directly, or more generally, an excitation voltage is applied with a drive electrode D and detected with a sense electrode S (Figure 9.22).

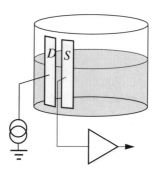

Figure 9.22   Liquid level sense

If the tank is conductive, the electrodes are applied to the inside or a small-diameter nonconductive manometer tube is used outside the tank. Alternatively, a single electrode can be used and its capacity to ground measured, or a probe immersed in the fluid.

9.10.1   Dielectric liquids

The total electrode capacitance for dielectric liquids is the sum of the electrode capacitance in air plus the variable capacitance. The variable capacitance varies as the percent of the electrodes immersed in the liquid times the dielectric constant of the liquid and increases with liquid level. The circuit measures the ratio of the capacitance between the plates and the amplifier capacitance to ground as shown in the equivalent circuit (Figure 9.23) with *Rgnd* infinite. The sense plate should be guarded on the back side as shown previously to minimize the effects of stray capacitance and pickup of unwanted signals.

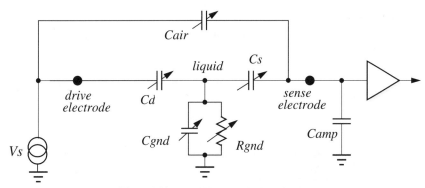

Figure 9.23   Liquid level equivalent circuit

9.10.2 Conductive liquids

For conductive liquids, the electrodes must be insulated or applied outside an insulating tank, and the capacitance may increase or decrease with liquid level. Two different effects are seen with conductive liquid sensors, or when sensing any conductive object. This is explained by considering the object's capacity to ground and drawing an equivalent circuit.

- *Cair* is the total capacitance through air paths with no liquid. It can be minimized by spacing the electrodes well apart, and by shielding air paths with a ground electrode.
- *Cd* is the variable capacitance between drive electrode and liquid; *Cs* is the variable capacitance between the liquid and the sense electrode. These are the capacitances to be measured.
- *Cgnd, Rgnd* is the impedance coupling the liquid to ground. *Rgnd* represents the resistivity of the liquid and often can be ignored.
- *Camp* is the amplifier input capacitance.

Several other variables such as dielectric leakage have been ignored.

The capacitive component *Cgnd* can be fairly large. The capacitance of an isolated sphere with diameter *d* in air

$$C = 2\pi\varepsilon d = 55.6 \cdot 10^{-12} d \quad \text{f, m}$$

is 55.6 pF for a 1 m diameter. Measuring *Cd* with the shunting effect of *Cgnd* is difficult. Guarding the entire tank to eliminate *Cgnd* would work, but is impractical. One easy approach is to use small electrodes and consider *Cgnd* large with respect to *Cd* and *Cs*. For this case, the circuit above does not work, as *Cgnd* would totally shunt the signal, so the level must be measured by measuring the capacitance of the drive electrode to ground as in

Figure 9.21. This can provide a solution for systems where approximate results are acceptable, or where a computer can be used to calculate an error table automatically.

With intermediate ratios of *Cd, Cs,* and *Cgnd*, the increase of *Cgnd* with liquid level may cause a decrease in coupling which exactly tracks the increase due to dielectric constant, thereby producing an unusable zero output at any liquid level.

9.10.3 Low-Z amplifier

A circuit rearrangement removes the effect of *Cgnd* and provides a more linear output (Figure 9.24).

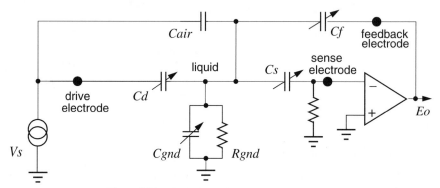

Figure 9.24   Liquid level with virtual ground amplifier

Here, a third feedback electrode is added. The voltage on the liquid is sensed through *Cs* and feeds the inverting input of the operational amplifier, so the liquid is at virtual ground and the effects of *Cgnd* and *Rgnd* can be ignored. Also, the variation of *Cs* and any stray capacitance at the amplifier input can be ignored. This produces a simpler equivalent circuit (Figure 9.25).

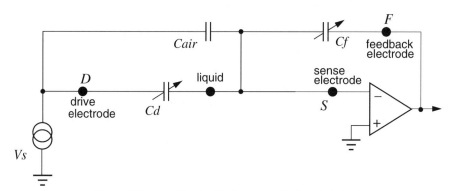

Figure 9.25   Liquid level with virtual ground, simplified circuit

With rectangular electrodes as above, the ratio of *Cf* to *Cd* is constant and the circuit output voltage does not change with liquid level. Changing the electrodes to triangle-shaped electrodes fixes this (Figure 9.26).

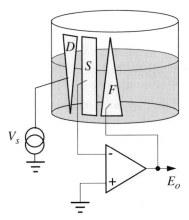

Figure 9.26   Electrodes for virtual ground liquid level sensing

The output voltage is then

$$E_o \; = \; V_S \frac{C_{DS}}{C_{FS}}$$

and will increase from near zero to *Vs* as liquid level increases.

9.11   LIQUID LEVEL SWITCH

A simple but accurate circuit generates an alarm signal when liquid level falls below a pre-set level (Figure 9.27).

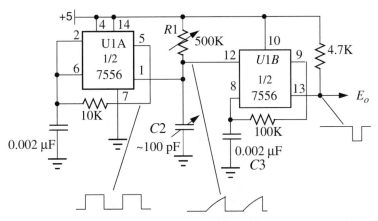

Figure 9.27   Liquid level switch

U1A is a square-wave oscillator at approximately 36 kHz, using the dual CMOS version of the popular 555 timer. *C2* is the capacitor to be measured, and can be the electrodes of Figure 9.22 except with the excited electrode grounded. During the 555's negative output

*C*2 is discharged; during the positive output *C*2 is allowed to charge towards +5 with a time constant of about 15 μs, adjusted by *C*2 and the potentiometer. When the liquid level falls below its set point, as adjusted by *R*1, the peak voltage at *C*2 rises until *U*1*B* is triggered at 2/3 of the 5 V supply. *U*1*B* produces a train of interrupt pulses at about 3.6 kHz as long as the liquid level is below the set point.

The timer can be a standard bipolar NE/SE556 or a CMOS ICM7556. The CMOS device will be more accurate as it has less input current and a more stable output voltage if impedances are kept high.

This circuit can also be used to build a single-plate proximity switch.

9.12   SIX-AXIS TRANSDUCER

A simple capacitive transducer is capable of measuring three axes of translation and three axes of rotation. The plate pattern is etched on two parallel printed circuit boards (Figure 9.28).

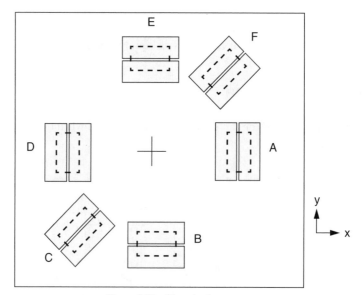

Figure 9.28   Six-axis plate pattern

The six driven electrode pairs shown shaded are on the bottom PC board and the six pickup electrodes shown dotted are on the top PC board. The bottom electrodes can be excited, for example, with 0° and 180° square wave signals at 50 kHz. Clockwise rotation around a perpendicular through the page is sensed by area variation of the pairs *F* + *C*, that is, the *F* pickup will pick up more of the plate on the right while the *C* pickup picks up more of the plate to its left. If the two pickups are exactly balanced, the sum will be sensitive only to rotation. Similarly, area variation of *A* + *D* handles translation in the *x* axis, area variation of *E* + *B* handles translation in the *y* axis, spacing variation of *E* − *B* handles rotation around the x axis and spacing variation of $A - D$ handles rotation around the y axis. Spacing variation of $A + B + D + E$ measures board separation.

The electronic circuit can be simple, as only one pickup amplifier is needed with various analog switches used to connect the pickup electrodes as area-variation-linear or spacing-variation-linear for the different measurements, and also to reconnect the drive electrodes to arrange for addition or subtraction as required.

9.12.1 Demodulator

Three axes of motion for the six-axis transducer are spacing variation sensors and three are area variation, but a single amplifier can be configured to linearize both.

Spacing variation

For a linear-with-spacing demodulator, if all eight drive electrodes in groups A, B, C, and D are connected together and all four associated pickups are connected together, the resulting capacitor can be connected as $C2$ in Figure 4.7. A reference capacitor is connected as $C1$ and is excited with, say, $0°$ at 50 kHz. The usual synchronous demodulator converts the output amplitude to DC.

Area variation

The linear-with-area circuit of Figure 10.26 can be used to demodulate the three area-variation motion axes. The driven electrodes are connected as $C1$ and $C2$.

Another method is to use the oscillator circuit of Figure 4.2 which can be used as both an area- and spacing-linear demodulator. Note that the analog switching circuits will contribute a large capacitance to ground, more than 80 pF for a 74HC4051 analog 8-input multiplexer. A virtual-ground amplifier with high bandwidth as in Figure 10.26 will handle 1 pF signal levels despite 80 pF stray capacitance, but canceling analog switch capacitance with the circuit of Figure 4.3 would require bootstrapping the *Vee* and *Vcc* pins of the analog switch.

9.13   CLINOMETERS

Clinometers measure inclination using the gravity vector as a reference. One model is discussed in Chapter 21; another with similar specifications is available from Lucas Control System Products, 1000 Lucas Way, Hampton, VA 23666, (800)745-8004. Lucas makes a clinometer using capacitive plates, a conductive liquid, and an inert gas, demodulated by two timers and providing analog ratiometric, PWM, and RS-232 output formats. Small-quantity prices are below $200. The specifications of a packaged version, DP-60, with a four-digit LCD display include a range of $± 60°$, resolution of $0.01°$, and cross axis error of less than 1% in a case $38 \times 138 \times 86$ mm. Another clinometer with a similar capacitive sensor is available from Lucas without a display and comes in a 50 mm round by 29 mm high enclosure. It offers $0.001°$ resolution and $0.1°$ linearity at small angles; the small quantity price is $120. Lucas can tailor the output curve to nonlinear functions such as sine and cosine by changing the electrode shape.

Tiltmeter

A clinometer with a small linear range, called a tiltmeter (Figure 9.29), has been described by Jones [1973]. Jones, as usual, explores the limits of precision available with very careful design. The instrument on a stable platform is capable of measuring 10^{-8} degree tilt, but the instability of the earth's crust is $10-100\times$ larger than the limiting sensitivity. As the frequency of these microseism tremors is in the 3–8 s range, a lowpass filter should be used to improve the measurement capability. Comparison of two instruments has shown that drift caused by mechanical instability is about one part in 10^9/day.

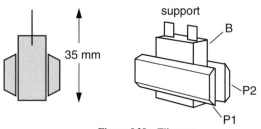

Figure 9.29   Tiltmeter

The two fixed electrodes $P1$ and $P2$ are adjusted with a narrow 75 μm gap to the weight B, which is suspended by two vertical strips. The size of the gap is calculated to provide a viscous damping force as air is squeezed from between the plates with motion, so that the expected resonance peak is critically damped. A 3 V rms excitation at 16 kHz is used with a 0°–180° balanced bridge circuit. The springs can be constructed from fused silica or single-crystal silicon for good stability and low hysteresis, and 70-30 brass and mica, with excellent stability, are used for the electrodes and for insulation.

10

Circuits and components

Circuit basics were covered in block-diagram form in Chapter 4. This chapter presents more design details, including schematics and component specifications. Capacitor dielectric properties are discussed in detail.

10.1 CAPACITOR DIELECTRICS

Considerable research on capacitor dielectrics was done at the Laboratory for Insulation Research at MIT in the 1950s and 1960s, and it is reported in a number of publications. In 1954, von Hippel discusses macroscopic and molecular properties of dielectrics, analyzes several dielectric applications, and presents an extensive list of dielectric parameters for many materials which has been condensed to Appendix 2. This laboratory also has published research into the dielectric properties of biological materials, materials at high temperatures, and measurement techniques.

The *IEEE Transactions on Dielectrics and Electrical Insulation* reports on recent developments in dielectric research.

10.1.1 Classical dielectric models

A lossy capacitor is classically modeled as a lossless capacitor with either a series or shunt resistor representing the loss term (Figure 10.1).

The series parasitic inductance L and the series resistance Rs can usually be ignored for our low current applications, or Rs can be converted to Rp for a particular frequency using

$$Rp = \frac{Xc^2}{Rs} = \frac{1}{(2\pi f C)^2 Rs} \qquad 10.1$$

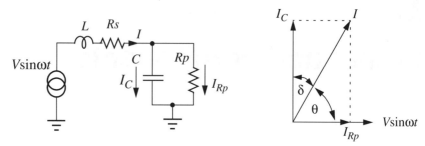

Figure 10.1   Lossy capacitor

The phasor diagram above (with *Rs* and *L* ignored) shows the relationship of the applied AC voltage to the in-phase current through the resistance and the 90° current through the capacitor. The capacitor *dissipation factor D*, also called *loss tangent,* tan δ, is the ratio of capacitive reactance X_C to resistance defined as

$$D = \frac{1}{Q} = \tan\delta = \frac{Rs}{X_C} = \frac{X_C}{Rp} = \frac{1}{\omega Rp C} \qquad 10.2$$

As frequency increases, I_C increases while I_{Rp} remains constant, so *D* decreases with increasing frequency if *Rp* is constant.

A similar dimensionless figure of merit is the power factor, defined as

$$PF = \frac{\text{power dissipation}}{\text{volts} \cdot \text{amperes}} \qquad 10.3$$

For *Q* > 10, *PF* is close to *D*.

The value of the equivalent lossless capacitor *C* is usually relatively independent of frequency over a wide range, but the behavior of the loss component *Rp* is not as simple. Various mechanisms contribute to loss, including actual conductivity due to migrating charge carriers and friction associated with the reorientation of polar molecules. Most commonly used dielectrics have a loss tangent which is relatively independent of frequency, implying that *Rp* decreases directly with an increase in ω. Figure 10.2, with data taken at various temperatures [von Hippel, 1954], shows that with the exception of Steatite ceramic (uncommonly used for capacitor dielectrics), loss tangent is remarkably constant with frequency. The loss tangent of the classic *RC* model with constant *Rp* is also plotted. The loss tangent remains relatively and remarkably constant through 16 frequency decades, 10^{-4} Hz–10^{12} Hz, for common capacitor dielectrics.

10.1.2 Modified dielectric model

The classic capacitor model (Section 10.2.1) is valid for a single frequency, but it does not account for the constant loss tangent with frequency. It also does not model two additional capacitor characteristics which can be observed empirically, dielectric absorption and insulation resistance change (or, equivalently, leakage current increase) vs. time.

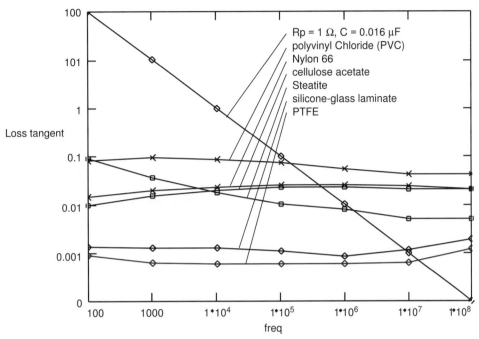

Figure 10.2   Loss tangent of various materials vs. frequency [von Hippel, 1954]

Dielectric absorption

Dielectric absorption is the excess charge absorbed, ignoring the normal charge expected from the equation $Q = CV$, by a capacitor polarized with a DC voltage. It is measured by charging a capacitor for 1 h at rated DC voltage and discharging through a 5 Ω resistor for 10 s. When the resistor is disconnected and the capacitor is open circuited, a percentage of the impressed voltage returns. The magnitude of this voltage is measured 15 min later. Some typical values of dielectric absorption are shown in Table 10.1.

Table 10.1 Dielectric absorption [Excerpted from (Westerlund). Copyright ©1994 by the Institute of Electrical and Electronics Engineers.]

	Dielectric absorption	tan δ	RC, seconds	n
Metalized paper	10%	0.028	3100	0.9821
PET (polyethylene-terephthalate)	1%	0.0025	38,000	0.9984
Polypropylene	0.003%	0.000075	1,250,000	0.999952

Table 10.1 also lists the self-discharge time constant (the product of the insulation resistance and the capacitance, RC) and a constant n which is related to the quality of the dielectric, as shown below.

Empirical model

Westerlund [1994] presents an empirical model of capacitor behavior which models the observed variation of R with frequency, and also models the decrease in capacitor leakage current vs. time. This model replaces C in Figure 10.1 with C_0, with C_0 very nearly equal to C for good dielectrics, and replaces Rp with

$$R_n \approx \frac{h_1(1-n)}{\cos\left(\dfrac{\pi n}{2}\right)\omega^n} \qquad\qquad 10.4$$

where h_1 is a constant related to capacitance. Westerlund's model closely tracks the observed charge absorption and insulation resistance change of common capacitor dielectrics, and it is considerably more accurate at predicting effects like the variation of leakage current vs. time and charge absorption than the standard RC capacitor model.

Insulation resistance vs. time

A constant DC voltage V applied to a capacitor C at $t = 0$ will produce a leakage current, in excess of normal capacitor charging current, of

$$i(t) \approx \frac{VC}{\left(\dfrac{1}{1-n}\right)t^n} \qquad\qquad 10.5$$

For a very good capacitor dielectric such as polypropylene, n (from Table 10.1) is close to 1 and i will be near zero. This equation implies that the insulation resistance R of a capacitor varies with time as t^n after applying a DC voltage.

10.1.3 Other dielectrics

Figure 10.2 shows that for a selection of miscellaneous plastics and ceramics the loss tangent is relatively constant vs. frequency. In most of these materials, the dielectric constant varies by less than 2:1. In contrast with the stable loss tangent of capacitor dielectric materials, the loss tangents of the miscellaneous materials shown in Figure 10.3 vary by factors of 100. Especially active is water, which is also responsible for the large change in loss tangent of materials like leather and paper for different relative humidities.

The relative dielectric constant of water is very high at about 80, and shows large changes with temperature; in Figure 10.4, water is shown with an expanded frequency range (note the nonlinear x axis). Measurement of water below 100 kHz is difficult because the angle of the loss tangent is nearly 90°. At 100 kHz, the loss tangent of 4 means that the impedance has an angle of 76°; the resistive component has a much lower impedance than the reactive component.

Dielectric measurement (Figure 10.5) has been suggested for the measurement of biological material properties [von Hippel, Oct. 1954], such as changes in cell permeability and cell content. Generally the loss tangent is a more sensitive indicator than dielectric constant. Aviation gasoline tested in Appendix 2 shows a change in loss tangent at 10 kHz from 0.0001 to 0.0004 as octane decreases from 100 to 91, while the dielectric constant change is from 1.94 to 1.95.

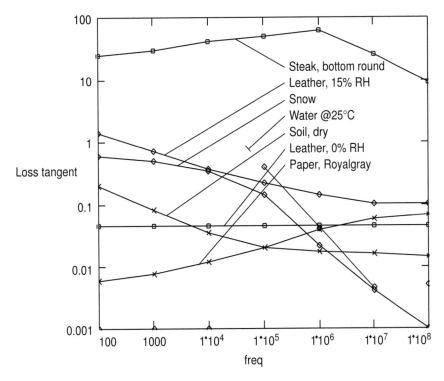

Figure 10.3   Loss tangent vs. frequency [von Hippel, 1954]

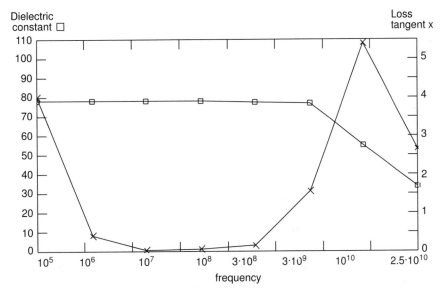

Figure 10.4   Dielectric constant and loss tangent of water vs. frequency at 25°C

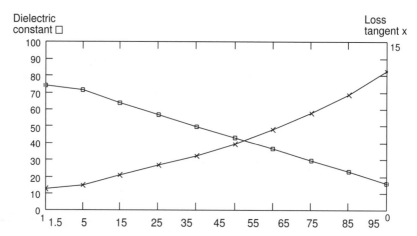

Figure 10.5   Dielectric constant and loss tangent of water vs. temperature at 1 MHz [von Hippel, 1954]

Another behavior of dielectric molecules is shown in this chart of Neoprene rubber (Figure 10.6). Polar molecules such as water and Neoprene often produce a temperature-sensitive peak in loss tangent, caused by a molecular resonance effect. As temperature is increased, the peak frequency will increase. The change of loss tangent may also be accompanied by a change in dielectric constant.

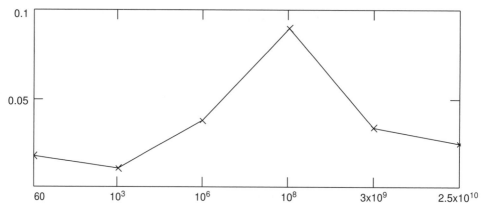

Figure 10.6   Neoprene loss tangent vs. frequency

10.2   CAPACITOR MEASUREMENT

10.2.1   Capacitor models

For low frequencies, two models are possible: the series model and the parallel model. Either model accurately represents the capacitance and the loss tangent of a capacitor at any single frequency. The parallel model is preferred, however, as the leakage current at DC is represented, and also if the dielectric is changed, the calculation of the new dielectric constant is simple, as the ratio of dielectric constants changes directly with the change in Cp. At higher frequencies three-terminal and four-terminal models including R, L, and C are used [East, 1964] to model the more complicated behavior of polar molecules.

Series

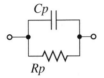

Parallel

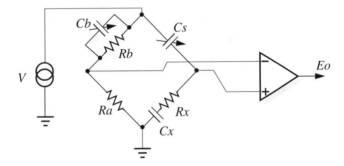

10.2.2 Bridges

The Schering bridge (Figure 10.7) is used for measuring capacitor and dielectric parameters [*Reference Data for Engineers*, p. 12-5].

Figure 10.7   Schering bridge

The capacitor under test is $CxRx$, and the amplifier is usually an instrumentation amplifier rather than an operational amplifier so the gain is stable and well defined. The following calculation is used when the bridge is adjusted to null.

$$Cx = Cs \cdot \frac{Rb}{Ra}$$

$$\frac{1}{Qx} = \omega CxRx = \omega CbRb$$

If a parallel model for the capacitor is needed, a modified Schering bridge is used, an example of which is shown in Figure 10.8.

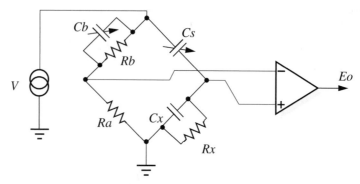

Figure 10.8   Modified Schering bridge

A carefully constructed Schering bridge for dielectric property measurements [Charles, 1966] has a loss angle sensitivity of 10^{-6} rad at frequencies of $1-10^5$ Hz.

10.3   PICKUP AMPLIFIERS

For capacitive sense systems with moderate resolution requirements, the pickup amplifier can be replaced by a simple comparator, or circuits such as an oscillator or a one-shot can be used. For circuits needing more accuracy a synchronous demodulator is needed, and the pickup amplifier needs higher precision and should be carefully matched to the application.

For good performance, the amplifier should have a high impedance input with low voltage noise and current noise at the excitation frequency. The circuit configuration may be a high-Z type, where the amplifier is used as a follower and has very high input impedance, or a low-Z type, where the amplifier's inverting input terminal is used as an input and heavy feedback results in low input impedance. A third option includes the sensor in the feedback loop and has some characteristics of each. These circuits are summarized in Chapter 3 and more fully discussed here.

10.3.1 High-Z amplifier

The guarded high-Z amplifier (Figure 10.9) is similar to circuits shown in Chapter 4, with an operational amplifier driving a guard to reject stray capacitance and to provide a shield against ambient electrostatic fields. Although this circuit is adequate for sensors with high sensor capacitance, above roughly 50 pF, it will be found unacceptable for smaller sensor capacitance because of amplifier parasitic input capacitance. In the figure above an additional bootstrap connection through Cv is brought to the Vee terminal of the amplifier. This connection nulls out the amplifier's internal capacitance from Vee to the input terminal and can improve the amplifier common-mode input capacitance from 5–10 pF to a fraction of a pF. Some electrometer-type amplifiers such as the AD549 and OPA129 (see Table 10.2) have 1 pF of common mode input capacitance and may not need Vee bootstrapping.

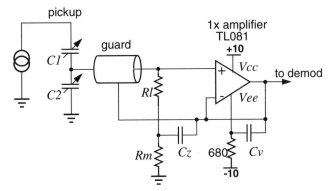

Figure 10.9 Guarded high-Z amplifier

Input resistance

The input resistance, *Rl* in Figure 10.9, is used to provide a path for amplifier input bias current and to set the input voltage to near zero, but it tends to shunt the signal and cause a spacing-dependent loss of output voltage with low-capacitance sensors. Figure 10.9 shows a way to bootstrap input resistance to higher values. The load resistor *Rl* is bootstrapped through *Cz* to the amplifier's output. With *Cz* large, this increases the effective value of *Rl* by a factor equal to the amplifier open-loop gain.

With a signal capacitance of 1 pF and a clock frequency of 20 kHz, the input time constant without *Cz* is *Rl* × 1 pF. As the time constant should be about 1/10 of the clock period for good waveshape, an input resistance of 500 MΩ is needed which is not only difficult to purchase, but can mandate a strenuous design effort to minimize stray currents. Strays caused by board leakage and op amp input bias current can affect gain and offset. With bootstrapping, a more convenient resistor value such as 10 MΩ can be used, and no problems will be encountered with stray current due to op amp input current or printed-circuit-board surface conductivity.

The value of *Cz* must be carefully chosen, as the bootstrap circuit may oscillate with some values. A SPICE simulation to check stability is recommended (Figure 10.10).

```
.define C1 10u
.define R1 22meg
.define R2 1k
```

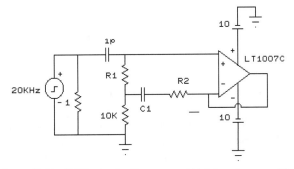

Figure 10.10   SPICE circuit of bootstrapped high impedance amplifier

These results were calculated (Figure 10.11). With low values of $R2$, considerable peaking is produced at 100 Hz which will usually cause the circuit to misbehave. As $R2$ is increased, this peaking disappears, but the bootstrap effect decreases; with $R2 = 100$, the gain peaking is minimal, but the bootstrap amplification of input impedance is reduced to 10×, or 220 MΩ. Bootstrapping can be useful for moderate increases of input resistance, but for high input impedance without peaking, start with the largest possible value of $R1$ and use other measures as described in Chapter 13 to interrupt leakage current.

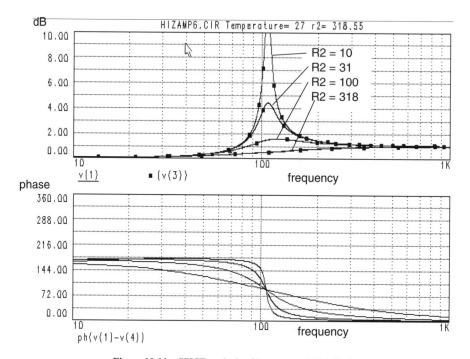

Figure 10.11   SPICE analysis of bootstrapped high-Z circuit

Stray capacitance

Several parasitic capacitances are found in an operational amplifier, including:

1. Capacitance between + and - inputs, which can often be found on the amplifier's spec sheet and is typically a few pF
2. Capacitance between + input and V - input
3. Capacitance between + input and V+ input

Capacitances 2 and 3 are not listed on the typical spec sheet, and they are not included in SPICE models. Some recent amplifiers do specify these capacitances; Texas Instruments' TLE2071 lists common mode input capacitance as 11 pF and differential-mode input capacitance as 2.5 pF. These strays can be handled in different ways:

- Capacitance 1 is not a concern as the negative input tracks the positive input and no current can flow in this capacitance, except if it is much larger than the sensor capacitance the output noise and rise time increase
- Capacitance 2 can be bootstrapped out by use of Cv as shown above
- Capacitance 3 cannot be bootstrapped in the typical integrated amplifier, as the $V+$ input is usually used as a gain stage reference voltage, and attempting to bootstrap it ruins the amplifier's gain

Parasitic capacitance simulation

Differential- and common-mode parasitic op amp capacitances are not included in SPICE operational amplifier models. If the typical 2 pF parasitic capacitances are added externally, the SPICE model of a high-Z amplifier becomes as shown in Figure 10.12.

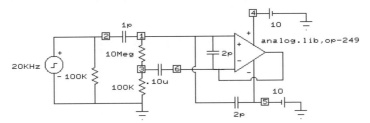

Figure 10.12 High-Z amplifier with parasitic capacitance

This amplifier shown is driven with a small, 1 pF sensor capacitance. With the attenuation of the parasitic capacitance, the output will have low amplitude and it will be very sensitive to the exact value of sensor capacitance. This may contribute an unwanted spacing sensitivity. The response with a 5 V square wave excitation is shown in Figure 10.13. The 2 pF capacitance to *Vee* is attenuating the signal by 3×.

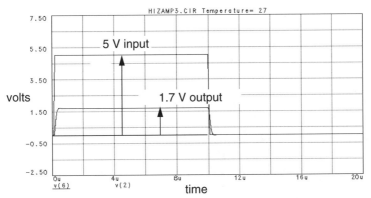

Figure 10.13 High-Z amplifier transient response

Adding a capacitor to bootstrap the *Vee* rail of the op amp as Figure 10.9 produces the SPICE circuit shown in Figure 10.14, featuring a much more accurate AC gain.

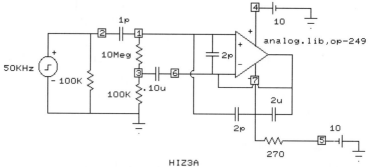

HIZ3A

Figure 10.14   High-Z amplifier, bootstrapped *Vee*

The AC response can be tested for peaking. The component values shown in Figures 10.15 and 10.16 produce a bridged-T notch response which will not impair circuit operation.

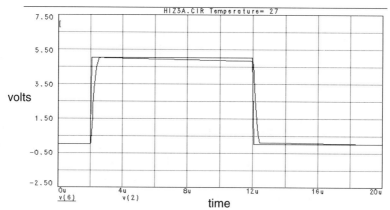

Figure 10.15   High-Z amplifier, bootstrapped *Vee*, transient response

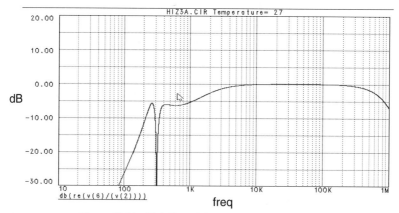

Figure 10.16   High-Z amplifier, bootstrapped *Vee*, AC response

Stabilizing the bootstrap connection

There is a hazard when bootstrapping V-. The positive feedback will make many amplifiers unstable (although not the TL081 shown), and cause oscillation at a frequency near the amplifier's f_T. This can be fixed by adding a lowpass filter to the bootstrap with a break frequency between the amplifier f_T and the frequency of the capacitive sensor drive (Figure 10.17).

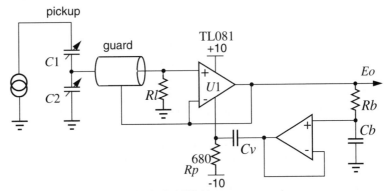

Figure 10.17   Stabilized bootstrap connection

The 3 dB frequency of the lowpass filter Rb, Cb should be about 1/4 – 1/10 of the amplifier's f_T. Rp is chosen to drop 2 or 3 V at the chosen amplifier's supply current, and Cv is chosen for an $RvCv$ time constant considerably longer than the period of the clock. A single op amp should be used for $U1$ rather than a dual or quad to keep total supply current low. This circuit works well with sensor capacitance greater than about 5 pF. Rise and fall times are about 5 μs, so signal clock frequency should be 10–50 kHz.

10.3.2   Rise time effects

Rise time change with spacing

As plate spacing increases and the sensor capacitance drops, the output rise time decreases proportionally. This is illustrated by the equivalent circuit (Figure 10.18).

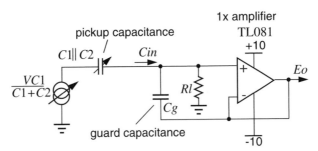

Figure 10.18   Rise time analysis

If the closed-loop amplifier gain is 1, Cg is fully bootstrapped out and Cin is zero or perhaps 1–2 pF of parasitic capacitance. If the closed-loop amplifier gain drops to K, however, the input capacitance increases to $Cg \times (1 - K)$. If, for example, the pickup plate is formed as an etch pattern on the top of a 0.062 in glass-epoxy printed circuit board (dielectric constant = 6) and the guard is formed with a similar etch pattern on the bottom, and assuming the signal electrodes were air-spaced at 0.062 in, Cg would be six times the pickup capacitance, $C1 \parallel C2$. If a signal frequency of 25 kHz and an amplifier with 2.5 MHz f_T and 6 dB/octave compensation is used, the amplifier gain at 25 kHz would be 2.5 MHz/25 kHz or 100 and closed loop gain drops to 0.99. The signal fundamental would be attenuated by 1%. Harmonics would be attenuated at a 6 dB per octave rate, so a square waveform will become somewhat rounded. As the spacing is further increased, the fundamental at 25 kHz will be more severely attenuated. To minimize this problem, maximize pickup capacitance, choose an op amp with high f_T and minimize Cg.

Slew rate limit

Another effect is the amplifier slew rate limit tabulated in Table 10.2. The high frequency reduction shown is a gain change only, but slew rate effects are nonlinear, produce a temperature-dependent kink in the transfer function, and should be avoided by selecting a high slew rate amplifier, using sine wave or trapezoidal excitation, preceding the amplifier with a lowpass or bandpass filter, or using a demodulator circuit that samples the amplifier output after the voltage has stabilized.

10.3.3 Discrete amplifiers

For applications with small sensor capacitance needing a high frequency clock, say over 100 kHz, the operational amplifier can be replaced with a discrete JFET or MOSFET amplifier, with suitable bootstrapping to reduce stray capacitance. Discrete amplifiers have good performance at a sacrifice of DC output precision, but DC offset is not usually a significant problem. A compound follower circuit with a small-geometry MOSFET (SD213) has a broadband high input impedance (Figure 10.19),

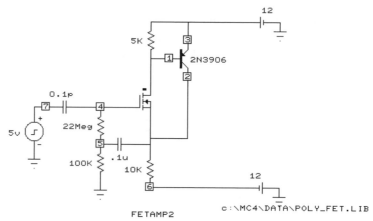

Figure 10.19   FET follower

with the FET gate-to-source capacitance, C_{GSS}, bootstrapped out by the follower connection. The remaining capacitance from gate to drain is smaller, about 0.2 pF, but will cause serious attenuation with the small 0.1 pF sensor capacitance illustrated. The response to a 5 V square wave shows that the input capacitance is about 0.5 pF (Figure 10.20).

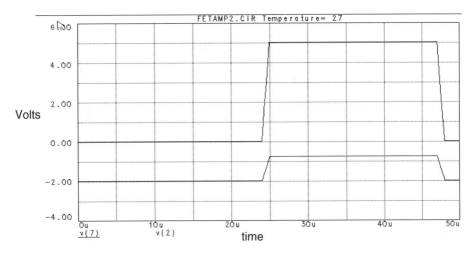

Figure 10.20   FET follower, transient response

A plot of AC gain vs. frequency shows the expected low frequency peaking caused by the input bootstrap (Figure 10.21). Although the *Vcc* rail cannot usually be bootstrapped with an integrated amplifier, the compound follower circuit can be bootstrapped to remove the effects of the FET gate-to-drain capacitance as shown in Figure 10.22, where *R*1-*C*1 help to minimize the amplitude of a 50 MHz high frequency peak. This circuit exhibits an input capacitance much smaller than the 0.1 pF sensor capacitance. The transient response shows about 95% amplitude which translates to a capacitance of 0.005 pF at the input (Figure 10.23).

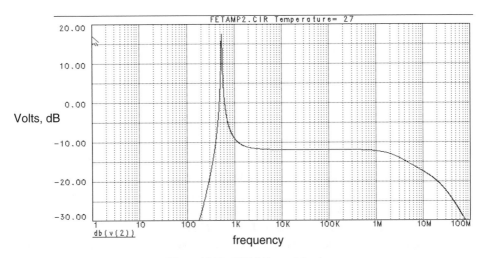

Figure 10.21   FET follower AC gain

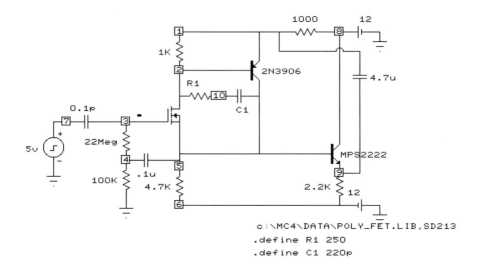

c:\MC4\DATA\POLY_FET.LIB,SD213
.define R1 250
.define C1 220p

Figure 10.22   Bootstrapped FET follower

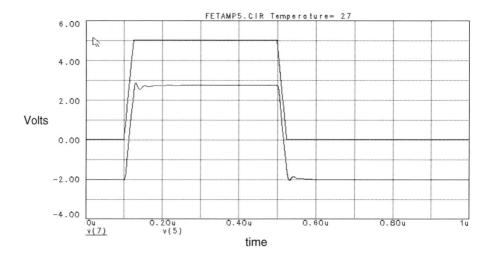

Figure 10.23   Bootstrapped FET amplifier transient response, 1 µs

10.3.4 Low-Z amplifier

Capacitive feedback can be used with a low impedance virtual-ground amplifier (Figure 10.24). This configuration has a feedback capacitance contributed by the coupling of the pickup plate to electrode *C* (Figure 10.25).

The capacitance between electrode *C* and the pickup plate produces the capacitive feedback, *Ccp*. Note that the effect of stray capacitance *Cstray* drops out if the gain of the amplifier is high.

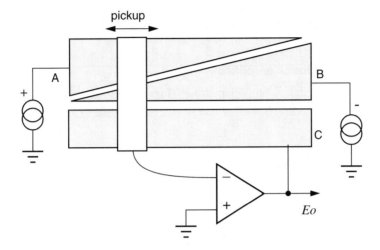

Figure 10.24   Low-Z amplifier with capacitive feedback

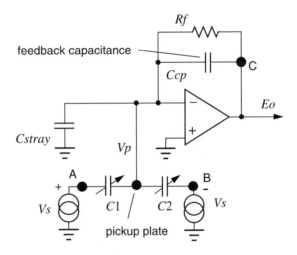

Figure 10.25   Equivalent circuit of low-Z amplifier

The amplifier output at frequencies well above the $R_f \times C1 \parallel C2$ time constant is

$$Vo \;=\; Vs \cdot \frac{C1 - C2}{Ccp} \qquad\qquad 10.6$$

As Ccp and $C1 + C2$ are usually all on the same substrate and share a common spacing, the capacitance ratio and therefore the gain do not change with plate spacing. This circuit is a very useful way to build spacing-independent sensors. Also, amplifier input capacitance is not as important with a low-Z amplifier (but output-to-input capacitance *is*); the high-Z amplifier circuits must use low-capacitance operational amplifiers.

Although the example above was shown with ramp-shaped plates, any of the more tilt-resistant geometries work, and multiple-plate arrays also will work with this configuration.

10.3.5 Feedback amplifier

The guarded high-Z amplifier circuit (Figure 10.9) can be modified by referring the ground to the amplifier input and floating the stator drive circuits (Figure 10.26).

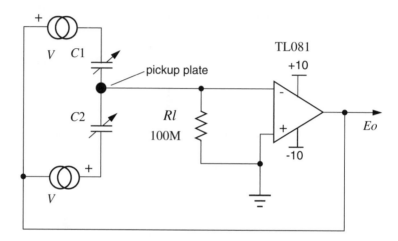

Figure 10.26   Feedback amplifier

With this circuit, the pickup plate is at virtual ground, so stray capacitance from pickup plate to ground is unimportant. This plate and its connection to the amplifier still need to be shielded with ground, however, as electrostatic pickup of noise sources is still a problem. The output voltage if $Xc \ll Rl$ is

$$\frac{E_o}{V} = \frac{C_1 - C_2}{C_1 + C_2}$$

10.7

This circuit performs well but suffers from the need to float the exciting voltage. A coupling transformer or a two-op-amp circuit or summing circuits as shown in Figure 4.8 can replace the floating drives.

T configuration

A high-value resistor Rf may be used across Ccp in Figure 10.25 to set the DC level at the amplifier output. The time constant $Rf \times Ccp$ should be much larger than the period of the clock to avoid tilt in the output waveform. For a 20 kHz system, a time constant of 250 μs can be used; then if Ccp is 1 pF, Rf should be 250 MΩ. Very large value resistors can be avoided (at the cost of greater output DC offset) by using smaller value resistors in a T configuration.

The T configuration (Figure 10.27) schematic shows the use of smaller value 1M resistors to simulate a 500 MΩ amplifier input resistance. This circuit has also a 500×

increase in DC offset voltage. If the increase in offset is a problem, a bypass capacitor *Cb* is added in series with the 2 kΩ resistor, with a value which produces an acceptable *Cb* × 2 kΩ time constant.

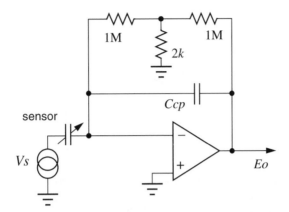

Figure 10.27   T configuration

10.3.6 Wireless connection

With one additional plate, Figure 10.24 can be made without wires. This works better with the low-Z connection of the op amp than with the high-Z connection (Figure 10.28).

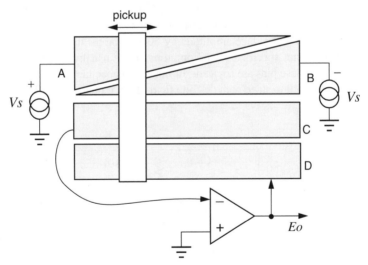

Figure 10.28   Low-Z, capacitive feedback without wires

Plate *D* has been added to Figure 10.28 to couple the amplifier's output back to the pickup plate without wires. For lower stray capacitance, a grounded plate should separate plates *B* and *C* and also *C* and *D*. The equivalent circuit is shown in Figure 10.29.

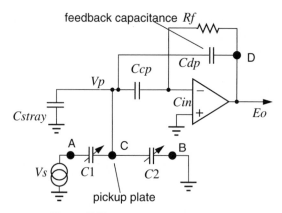

Figure 10.29 Wireless equivalent circuit

As in the previous circuit, a large value resistor Rf is included to set the DC output. The time constant $Rf \times Cdp$ should be several times larger than the period of the signal, or the T connection can be used to multiply Rf's value.

The amplifier gain is reduced by the attenuation of the input coupling capacitor Ccp looking into the amplifier differential mode input capacitance Cin, or about 2–10×. Including allowance for this effect, the amplifier f_T should be 20–100 times the oscillator frequency.

FET input amplifier

With low-Z amplifier circuits like Figure 10.28, different stray capacitances become important compared to high-Z amplifiers. Capacitance from amplifier input to output now reduces signal, but input capacitance and other strays are not important. 1 pF of input-to-output capacitance will reduce amplitude by 50% if the pickup capacitance is 1 pF. This stray capacitance is not specified for IC op amps and is usually small, except in a dual op amp pinout where these pins are adjacent. In-out stray capacitance can be further reduced by adding an input preamp stage with discrete field-effect transistors (FETs) (Figure 10.30).

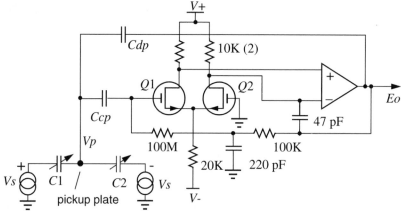

Figure 10.30 FET-input amplifier

The discrete FETs almost completely decouple the amplifier output from the input, and also improve AC parameters like noise, slew rate, and f_T, as the compromises of monolithic amplifier fabrication do not permit optimizing the input FET characteristics. This circuit can be constructed to have an infinitesimal value of input to output coupling, and because of this virtue it is suggested for use with extremely low values of sensor capacitance.

The 47 pF capacitor stabilizes the amplifier by reducing its 0-dB-gain frequency to compensate for the extra gain of $Q1$; in some cases a small (470 Ω–2 kΩ) resistor may also be needed in series with the source of $Q1$ to reduce the gain of this stage. $Q2$ can be deleted and the amplifier's negative input connected directly to a DC reference, say 2.5 V, if DC offset is not a problem. An advantage of this circuit is that a bipolar input op amp can be used, as the increased input current and current noise of bipolar amplifiers compared to FET versions are not important here.

10.4   NOISE PERFORMANCE

Current noise

With small values of pickup capacitance, low amplifier input noise current is needed. If, for example, a 1 pF pickup capacitance is used, with this sensor capacitance its impedance at a typical signal frequency of 25 kHz will be 6.4 MΩ. If the bipolar MC33077 were used, from Table 10.2, its current noise is 0.6 pA/$\sqrt{\text{Hz}}$. Notice that 0.6 pA is the noise measured in 1 Hz bandwidth; the noise increases as the square root of bandwidth. This current noise produces 18 pA of noise in a bandwidth of 1 kHz, for an unsatisfactory input-referred noise voltage of 120 μV rms.

Crossover resistance

The other amplifiers listed in the table are FET input types and will have lower current noise contribution, a significant advantage with pickup capacitance less than a few pF. The amplifier crossover resistance R_c is the equivalent source resistance at which the typical current noise i_n of the op amp becomes larger than the voltage noise e_n. Larger values of crossover resistance are better. For the National LMC6081A1, for example, with a circuit resistance (as viewed from the op amp input terminal) of 90 MΩ, the amplifier current noise and voltage noise contribute equally to the total output noise. The amplifier-noise figure of merit for a sensor with an impedance X_c is $e_n + R_c/X_c$. Smaller values are better.

For input impedances less than R_c, voltage noise predominates. An amplifier with an R_c greater than the impedance of the electrode system is desirable, although the amplifier voltage noise is also significant. As the voltage noise of most amplifiers falls in a fairly narrow range between 4 and 40 nV/$\sqrt{\text{Hz}}$, the noise at the amplifier output is usually minimized if R_c is larger than the circuit impedance.

Noise resistance

Another noise resistance R_n can be defined as

$$R_n \; = \; R_v + \frac{X_c^2}{R_i} \qquad\qquad 10.8$$

where R_v is a resistor value which generates a thermal noise of the same magnitude as the amplifier input voltage noise, R_i is a resistor value which generates a thermal current equal to the amplifier input current, and X_c is the impedance of the sensor capacitance. Smaller values of R_n are better. For FET transistors, R_n is on the order of 1 kΩ, improving for larger geometry devices.

Resistor thermal noise

The thermal noise in a resistor R (see also Section 12.1.1) is

$$E_t = \sqrt{4kTRB} \qquad\qquad 10.9$$

where

$4kT = 1.64 \times 10^{-20}$ W-s at 25°C

B = noise bandwidth, Hz

For a 1M resistor and 1 kHz noise bandwidth at 25°C, $E_t = 4.06\ \mu$V.

10.4.1 Circuit noise

Feedback circuit

The noise performance of the feedback circuit with a sensor capacitance of Ct and an exciting voltage of Vs can be calculated based on this equivalent circuit (Figure 10.31):

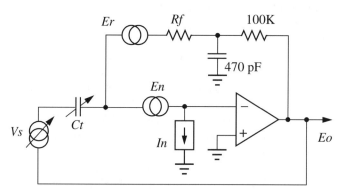

Figure 10.31   Noise model, feedback circuit

- Rf sets the DC output to zero
- En, In are the amplifier's equivalent input voltage and current noise generators
- Er is Rf's thermal noise
- The 100 kΩ / 470 pF lowpass filter (3 kHz) is included so that the input impedance is not reduced by Rf

With high op amp gain and high op amp input impedance, and assuming Rf is large, the output noise E_{no} is

$$E_{no}^2 = (E_o - V_s)^2 \cong \dot{E}n^2 + \left(\frac{I_n}{sC_t}\right)^2 + \left(\frac{E_r/Rf}{sC_t}\right)^2 \qquad 10.10$$

Output noise decreases with increasing operating frequency and increasing sensor capacitance, and output noise decreases with increasing Rf until $Rf = Rc$. This behavior is also true for the other amplifier configurations, and eq. 10.10 is similar for noninverting amplifiers with a gain of 1 or inverting amplifiers with a gain of 1. For the best noise performance with a given sensor capacitance Ct:

- Choose an amplifier with high Rc, greater than $1/\omega Ct$, and low En
- Maximize the excitation voltage Vs
- Increase Rf so it is greater than Rc, or replace it with a noiseless switch circuit to sample output DC back to the input
- Increase the excitation frequency so that the impedance of Ct is less than Rc

Under these conditions the output noise will be close to En, and the signal-to-noise ratio Eno/En will be Vs/En. Although sensor series resistance is not included, its effect is minimum. If this resistance is 10 kΩ, for instance, it will cause a noise voltage of 12 nV/$\sqrt{\text{Hz}}$, about the same as an op amp. Also, the stray capacitance to ground at the sensor should be less than the sensor capacitance. If $Cstray = Ct$, the noise will increase by 1.414.

Evaluating the output noise for the Texas Instruments CMOS-input operational amplifier TLE2061 with the following operating conditions:

$$\text{freq} = 20 \text{ kHz}$$
$$Rf > Rc = 100 \text{ M}\Omega$$
$$Er = 4 \text{ μV}$$
$$Ct = 1 \text{ pF}$$
$$Z_{Ct} = 1/\omega C = 8 \text{ M}\Omega$$
$$\Delta f = 100 \text{ Hz (noise measurement bandwidth)}$$

the output noise voltage is

$$Eno \cong 0.11 \cdot 10^{-6} \quad \text{V rms}$$

The additive noise is considerably less than 1 μV, even with the small sensor capacitance.

10.5   COMPONENTS

10.5.1 Operational amplifiers

Operational amplifiers suitable for capacitive preamps have a high input impedance, high slew rate, high f_T, low power supply current consumption, and low current noise. The optimum input resistance for minimum noise, R_c, equal to the noise voltage divided by the noise current, is also listed. Some examples are shown in Table 10.2.

Table 10.2 Low-current-noise operational amplifiers

Mfgr.	Type	f_T MHz typ	Slew rate V/μs typ	I_{bias} 25°C typ	I_{supply} mA per amp typ	Voltage noise nV/$\sqrt{\text{Hz}}$ 1 kHz	Current noise fA/$\sqrt{\text{Hz}}$ 1 kHz	Cross resist. R_c	Notes (JFET types except where noted)
Analog Devices	AD549	1	3	100 fA	0.6	35	0.11	318 M	$Zin=10^{13}\,\Omega$, 1 pF
Analog Devices	AD820/822	1.8	3	10 pA	0.6	16	0.8	20M	Single supply, rail-to-rail output, input CMR includes gnd
Analog Devices	AD823	16	20	3 pA	2.6	16	1	16M	Dual, 3 V, rail-to-rail
Burr-Brown	OPA129	1	2.5	±30 fA	1.2	27	0.1	270 M	$Zin=10^{13}\,\Omega$, 1 pF
Harris	CA5160	15	10	2 pA	2	30[a]	N/A	--	MOSFET[b]
Linear Tech	LT1169A	5.3	4.2	1 pA	5.3	6	0.8	7.5 M	Dual
Motorola	MC35081	8.0	25	60 pA	2.5	38	10	3.8 M	C_{in} = 5p
Motorola	MC33077	37	11	280 nA	1.75	25	600	42 k	Dual, bipolar
National	LMC6081	1.3	1.2	4 pA	0.85	--	0.2	90 M	CMOS[c]
National	LM6142	17	30	280 nA	0.88	16	220	72 k	1.8–24 V supply, rail-to-rail input and output, bipolar
National	LMC6482	1.5	0.6	0.02 pA	0.9	37	30	1.2 M	Dual; 3–15 V supply, rail-to-rail input and output
T.I.	TLE2061	1.8	3.4	3 pA	0.28	43	1	60 M	--
T.I.	TLE2082	10	35	15 pA	1.5	16	2.8	15 M	Dual
T.I.	TLC271	17	3.6	0.6 pA	0.9	25	--	--	@high bias; 3–16 V supply
T.I.	TL081	4.0	13	--	1.8	--	10	3 M	--

[a]At 10 kHz

[b]16 V max supply

[c]16 V max supply; common mode input range includes ground

The devices with bias current higher than 1 nA are bipolar-input types which are not directly useful for this application because of high current noise, but can be used if preceded with a discrete FET preamplifier. Other amplifiers, unless noted, are FET input types with one amplifier per package. The FET input amplifiers show much smaller bias currents and much lower current noise than the bipolar types. The supply current can be important in the follower connection where the negative supply is bootstrapped.

10.5.2 Resistors

High-value resistors

Most resistors are conveniently available in values ranging from 10 Ω to 10 MΩ. 100 MΩ resistors and over are a specialty item available from several suppliers, including Ohmtek, 2160 Liberty Drive, Niagara Falls, NY 14301, (716)283-4025 and Ohmcraft, Inc., 3800 Monroe Ave., Pittsford, NY 14534, (716)586-0823. These companies are subsidiaries of Vishay Resistive Systems of Malvern, PA. Ohmcraft's fine line thick film technology produces resistors with these specifications (Table 10.3).

Table 10.3 High value resistor specifications

Resistance	10M	100M	1000M	10,000M
Tightest tolerance	1%	5%	5%	10%
TCR, 25–75°C, ppm/°C	25	50	100	150
Voltage rating	500 V	500 V	1000 V	1000 V
VCR, ppm/V	<10	<10	<10	<10

Chip resistors are available in 0502 (5 × 2.5 mm) size with less than or equal to 100 GΩ resistance, and larger sizes such as the 2512 (25 mm × 12.5 mm) can have 3.5 TΩ resistance, a DC voltage rating of up to 2500 V, and 2 W power ratings.

10.6   CARRIER

Frequency of operation

As the operating frequency is increased, the impedance of the capacitive sensor drops. This is usually valuable, as for small sensor dimensions the capacitance may be a fraction of a pF. With 0.01 pF capacitance at 10 kHz

$$X_c = \frac{1}{2\pi f C} = 159 \text{ M}\Omega$$

At this impedance it will be difficult to obtain a good signal to noise ratio, with additive amplifier current noise and capacitive crosstalk being the main noise sources. With the frequency increased to 200 kHz, an amplifier with a sensor capacitance of 0.01 pF will have an output noise approaching the noise voltage of the amplifier and a signal-to-noise ratio with a 5 V excitation of about 170 dB.

But high frequency operation has two problems, power and ease of demodulation. Power increases because higher frequency amplifiers are needed, and demodulation is more difficult as sine wave signals must be used rather than square wave to avoid risetime induced nonlinearities. With a lower frequency carrier, say 40 kHz square wave, the amplifier bandwidth should be roughly 400 kHz to preserve good waveform shape, and CMOS switches can be conveniently used for demodulation. With a 1 MHz carrier, good square wave fidelity would mandate a 10 MHz amplifier and < 50 ns switches. These parts would have considerable power consumption.

Sine wave carrier

Sine wave sensor excitation can be used, but amplifier bandwidth should still be a large factor higher than the carrier frequency for good closed-loop gain stability, and sine wave demodulation will require high frequency multipliers and other expensive parts. In general, simpler circuits can be built in the 10 – 100 kHz range using operational amplifiers with 1 – 10 MHz bandwidth and CMOS switches of the 74HC4053 type which switch in the 100 ns range. For critical circuits, the sine wave carrier offers some benefits to offset the extra cost:

• A narrow bandwidth tuned circuit can precede the amplifier to attenuate out-of-band noise

• Small timing variations between the sensor excitation signal and the demodulator reference signal do not contribute to drift

• Spikes and noise are reduced

• A lower frequency amplifier can be used, and the amplifier slew rate is usually unimportant

Also, a sine wave drive is essential for capacitive micrometers or other circuits where the amplifier gain is increased in order to expand the range. For example, if plate deflection in a balanced bridge were only 1% of the total distance, the amplifier gain would be increased by 100× to produce a full scale output. In this case, the bridge is adjusted for null by tweaking an adjustable leg to balance the output to zero with no deflection. With square wave drive, the bridge must successfully balance not only the fundamental, but all of the square wave harmonics, or the amplifier at null will output excess harmonic energy. When the fundamental is completely nulled, it is likely that nonlinear effects will cause the higher harmonics to feed through, spoiling the null and possibly saturating the amplifier output. A tuned input circuit with passive components can remedy the problem, or sine wave drive may be more convenient.

Sine wave generation

Sine wave sensor excitation has advantages of avoiding slew rate limit and nonlinear phase shift problems. At high frequencies, sine waves produce less EMI and are preferred for applications which radiate energy such as proximity detectors. For sensitive applications, sine wave excitation or a bandpass immediately preceding the amplifier are suggested. Sine waves do not need to have particularly good distortion characteristics for sensor use; more important is amplitude stability; this tends to rule out traditional sine wave oscillator circuits such as Colpitts and Wein Bridge oscillators. Several techniques can be used to produce sine waves with timing derived from a quartz crystal oscillator and with very stable amplitude:

1. Square wave with lowpass filter
2. DAC with sine coefficients
3. Finite impulse response (FIR) digital filter

The square wave with lowpass filter is easiest to build, but it adds a phase shift which may not be stable. If the same square wave signal is used to synchronize the demodulator, the phase shift produced by the filter must be equalized in both paths for accurate demodulation. One method is to use a comparator to square up the filter-output sine wave for the demodulator.

Methods 2 and 3 can use a coarse 3 bit DAC or a very short three-tap FIR lowpass filter to produce sine waves with no 2nd through 4th harmonic distortion. Higher order harmonics may be ignored, or can be easily attenuated with a low-phase-shift one-pole lowpass filter.

Square wave generation

Square wave generation is simple with CMOS logic, as this logic pulls the high impedance of the typical capacitive sensor to the power rails very accurately. With a low-value resistive load, say less than 10 kΩ, a low impedance family such as the 74AC series can be used. See also Section 10.6.

10.7   DEMODULATORS

10.7.1 180° drive circuits

Two different types of square wave drive voltages are useful, 90° and 180°. The 180° driver is the simplest and is used for sensors which have a stable electrode spacing, or for systems with moderate total accuracy requirements (Figure 10.32).

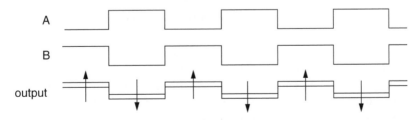

Figure 10.32   180° drive waveforms

These drive signals can be conveniently and accurately generated by CMOS logic gates, such as the 74HC04. With 5 V logic, as the pickup plate moves from A to B, the output varies from 0° (5 V p-p), through zero, to 180°. A ratiometric area-variation motion sensor with a 180° drive is shown in Figure 10.33.

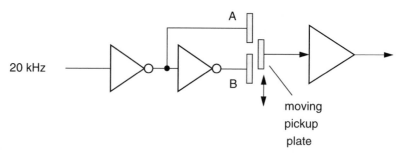

Figure 10.33   180° drive circuit

For simple circuits, only one plate needs to be driven, with the other plate at ground. Driving both plates gives 2× signal swing and produces a ground-centered output.

A commercial demodulator for 180° drives is available from Analog Devices, the AD698 (Figure 10.34). It is intended to be used as a demodulator for linear variable differential transformer (LVDT) sensors, but it is also used as a capacitive sensor demodulator.

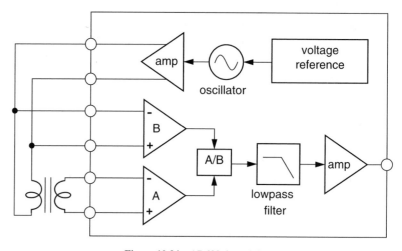

Figure 10.34   AD698 demodulator

The stabilized oscillator produces a low distortion 20 kHz sine wave drive with up to 24 V p-p rms with a ±15 V supply. The *B* channel input is a synchronous demodulator which detects the excitation voltage amplitude, and the *A* channel input is a second synchronous demodulator. The use of the *B* channel as a reference means that the oscillator amplitude stability is unimportant. A divider produces an accurate ratiometric output which is lowpass filtered and buffered, and the divider output is the ratio of the transformer output to the excitation voltage amplitude. The typical gain error is 0.1%, and typical nonlinearity is 75 ppm of full scale. The input impedance is 200 kΩ, as the device was designed for low-impedance inductive sensors, so a buffer must be used for most capacitive sensors (Figure 10.35). If reference capacitors with the same construction as the sense

capacitors and a second ampilfier are added to the *B* input, temperature effects and amplitude variations caused by spacing change may be compensated.

A similar device without the ratiometric feature and with less accuracy, Analog
Device's AD598, may also be used, or the AD2S93 may be used if its internal 14 bit ADC
is needed.

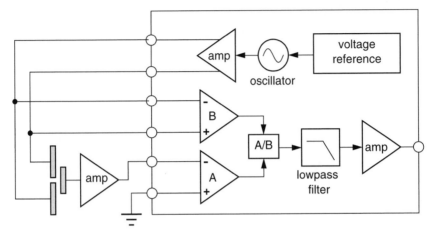

Figure 10.35   AD698 with bridge-type capacitive sensor

10.7.2 90° drive circuits

For improved performance with variable spacing, 0 and 90° excitation signals are needed
(Figure 10.36).

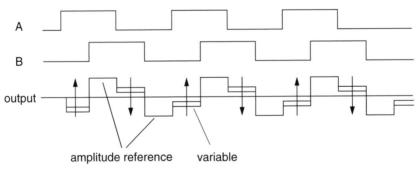

Figure 10.36   90° drive

With the 90° drive, the output has an amplitude reference, alternately reversing
polarity, which does not change amplitude with pickup plate position. Between amplitude
reference segments, the variable segments also alternately reverse polarity as shown, and
change amplitude in response to motion. This allows amplitude-independent ratiometric
demodulation which is useful for systems with poor mechanical gap tolerances or gap
variation with position; during each cycle the ratio of the variable segment to the fixed

amplitude reference is measured. The demodulator can accommodate large amplitude changes without compromising output accuracy.

Amplitude reference

The reference segments can be used in two ways: as an AGC signal, to control the gain of a variable-gain amplifier, or as a reference for an analog-to-digital converter or a voltage-to-period converter. In either case, the circuit output is the ratio of a variable segment to an amplitude reference segment. The graphic input tablet described in Chapter 19 shows how this circuit may be implemented.

11

Switched capacitor techniques

Capacitive transducer circuits which are built on conventional printed circuit boards can use standard analog design techniques, but designs for implementation on silicon are best done with switched capacitor circuits using analog discrete-time approaches, as accurate resistors are unavailable on silicon. MOS technology has become the most widely used silicon fabrication technique, suitable for large-scale digital circuits with 10^6 transistors as well as small application-specific ICs for mixed analog and digital functions. MOS transistors also make excellent switches, and switched capacitor implementations of analog circuits for capacitive sensors offer an excellent combination of accuracy, low production cost, and integration with digital and computer logic circuits. This chapter is an introduction to switched capacitor technology. For an in-depth treatment, read *Switched Capacitor Circuits* by Allen and Sánchez-Sinencio [1984].

11.1   ALTERNATE DESIGN TECHNIQUES

11.1.1   Digital signal processing

DSP (digital signal processing) techniques are becoming more popular as powerful DSP microcomputers are becoming available for low cost. With a DSP implementation, the low-level signal from the sensor electrodes is lowpass filtered, digitized to the required precision and at the required sampling rate, and input to a DSP computer for processing. The output digital stream is converted back to a continuous analog signal, if needed, with a DAC (digital-to-analog converter) and an output reconstruction filter. DSP techniques are accurate and capable of handling complex algorithms, but development time and power consumption are unfavorable, and more silicon area is usually needed.

If a local computer is needed for other purposes and has some unused cycles, the DSP method may be best. A very rough estimate of instructions-per-second compute power for capacitive sensor processing can be made:

- Assume the frequency response needed from the transducer is f_C
- Choose $10 f_C$ for the sensor clock f_S, except f_S should be greater than 10 kHz or so to keep capacitive reactances in a reasonable range
- Each half-cycle of f_S may need this processing:

    1. interrupt
    2. input sample value
    3. demodulate by inverting alternate half cycles
    4. filter by averaging both half-cycle values
    5. output sample at f_S rate
    6. return from interrupt

Assuming the desired computation precision is handled by a single CPU instruction, this calculation may require 20–30 instructions at a rate of $2f_S$. At 10 kHz, this is 0.4–0.6 MIPS (million instructions per second), well within the range of almost all processors. Adding other signal processing such as AGC or more complex filtering will double or triple the MIPS required.

DSP computers suited for this use are available from Analog Devices, ATT, Texas Instruments, and many other manufacturers, for prices starting at about $15. They handle 16 or 32 bit fixed point or floating point number crunching, and excel at the multiply-and-accumulate operations needed for digital signal processing. While a general purpose 8 bit microcomputer may be limited to 1–5 MIPS, DSP computers are 10–100 times faster and provide more precise math.

11.1.2 Charge coupled devices, bucket brigade devices

Other silicon implementations which can be considered are CCDs (charge coupled devices) and bucket brigade devices, but these can have a poor signal-to-noise ratio and are high volume solutions only, perhaps 1M units, as tooling expenses are high. Also, postfilter antialiasing requirements are more stringent as the sampling rate is closer to the maximum input frequency [Davis, 1979].

11.1.3 Packaged switched capacitor filters

Active filters built with switched capacitor circuits are available from several manufacturers. National's MF4, MF6, etc., are typical examples. They are packaged in 14–20 pin ICs and sell for less than $10. Their advantages are:

- Up to 12 pole filter complexity
- Input frequency range 0.1 Hz–100 kHz
- Frequency response proportional to clock frequency
- Frequency accuracy to ± 0.3% typ
- Dynamic range of 83 dB
- Low voltage, low current power requirements
- Gain accuracy ± 0.15%

These devices can build allpass, highpass, lowpass, bandpass, or band reject filters with manufacturer-selected or user-selected filter response choice. An advantage over continuous analog implementations is the ease and accuracy of tuning the response frequency by adjusting the clock. This is somewhat complicated by the necessity of providing an input antialias filter which will cover the range of clock tuning, but as the clock is usually 25 to 100 times the filter response frequency, the input antialias filter will be simple and may not need to be tuned. A disadvantage, common to all switched-capacitor circuits, is the limited frequency response. As the clock may be a high multiple of the maximum signal frequency, the signal is limited to the low hundreds of kHz.

Some of these devices are noisy compared to continuous-time realizations; this parameter should be checked carefully for sensitive applications.

11.2   COMPONENT ACCURACY

Capacitive sensor circuits use several types of components: operational amplifiers, switches, resistors, and capacitors. Of these components, the resistors and capacitors have the most effect on circuit accuracy, as amplifiers and CMOS switches are available which are quite accurate. With conventional PC board construction, accurate discrete components are available, but when converting a design to silicon, the designer must be aware of tolerance limitations. An *RC* circuit's frequency response varies as the product of two component types with poor absolute accuracy, while a switched capacitor filter with an accurate clock varies as the ratio of similar components, capacitors, so that absolute precision is unimportant and the ratio is controlled by very precise and repeatable photolithography. With silicon implementations, probably the worst case is for large value *RC* time constants which might be used in a low frequency active filter. As an example, for a 10 kHz filter, a resistor of 100 kΩ and a capacitor of 159 pF could be used. These two components would take up 500 mil^2 of silicon and have a tolerance of 50% and a temperature coefficient of almost 0.2%/°C. With a switched capacitor equivalent, the (ratio) tolerance improves by a factor of 500 to 0.1% and the temperature coefficient improves to better than 25 ppm/°C.

11.2.1 MOS vs. bipolar processes

Either MOS or bipolar processing can be used for capacitive sensor processing circuits. Bipolar processes produce superior high gain, high frequency bipolar transistors, while MOS (metal oxide semiconductors) processes make high input impedance transistors and very accurate switches. In general, for low power consumption and low frequency applications (below 1 MHz), MOS processes will be superior; for higher frequencies, bipolar transistors and g_m-C filters can be used. MOS (or JFET) transistors have considerably more 1/f noise than bipolar transistors which can be a negative factor for baseband systems, but capacitive sensors typically use an excitation frequency which is above the 2–20 kHz 1/f corner of low noise MOS transistors.

11.2.2 BiCMOS process

BiCMOS processes combine bipolar and CMOS transistors on the same chip, and have been developed for digital and memory applications, particularly to provide a low impedance, high current output buffer with small area. As the process moves into the mainstream, it has been adopted for improved analog circuit designs.

The most obvious improvement is the use of bipolar emitter-follower buffers for driving output pins or low impedance on-chip circuits. Bipolar transistors have a lower and more predictable voltage drop, smaller size, and higher bandwidth with capacitive loading.

The commonly used differential-cascode connection for operational amplifiers is improved if PMOS input transistors, contributing near-zero input current and lower $1/f$ noise than NMOS, are followed by a differential common-base bipolar n-p-n pair [Wooley, 1990]. A drawback for high impedance sensors is the increased size and increased parasitic capacitance of PMOS relative to NMOS transistors, so with an excitation frequency higher than the $1/f$ corner frequency of NMOS, NMOS may be preferred. Either polarity of MOS is preferable to bipolar for high slew rate and low input current, as input-stage bias current can be increased for MOS to increase input-stage slew rate (generally the limiting slew rate for the complete amplifier) without the accompanying penalty of higher input bias current found with bipolar transistors.

Other circuits which are improved with BiCMOS are reference voltages (band gap references) and temperature-dependent voltage generation using the well-known accurate temperature dependence of the forward voltage drop of similar diodes with dissimilar forward current. These circuits can be also built with a small performance degradation using standard CMOS processing, as the low gain, low frequency response parasitic bipolar substrate transistors can be pressed into service as diodes.

11.2.3 MOS vs. discrete component accuracy

Tables 11.1 and 11.2 compare the accuracy of MOS components with discrete components. The discrete devices were chosen to represent the highest available precision.

Table 11.1 Component accuracy, NMOS [Allen et al., p. 560; Colclaser, p. 237]

Component	Relative accuracy	Absolute accuracy	Voltage coef., ppm/V	Temp. coef., ppm/°C	Range of values
Poly-to-poly capacitor	0.1%	3%	−30	+25	0.15–0.2 pF/mil^2
MOS capacitor	0.1% (10 μm)	3%	−20	+25	0.25–0.3 pF/mil^2
Diffused resistor	2% (5 μm) 0.23% (50 μm)	±50%	−200	+1500	10±2 Ω/□
Polysilicon resistor	2% (5 μm)	±50%	--	+1500	60 Ω/□
Ion implanted resistor	2% (5 μm) 0.15% (50 μm)	--	−800	+400	500–20 kΩ/□
FET resistor	4%	±25%	Voltage dependent	--	10 k–100 kΩ

Table 11.2 Component accuracy, discrete [Mazda, pp. 12-19, 13-8; also various manufacturers' data sheets]

Component	Relative accuracy	Absolute accuracy	Voltage coef., ppm/V	Temp. coef., ppm/°C	Range of values
Resistor, carbon film	1%	1%	250	±500	1 Ω – 1 MΩ
Resistor, metal film	0.1%	0.1%	0	±5	10 Ω – 1 MΩ
Resistor, wire-wound, Cu/Ni	0.05%	0.05%	0	+20	1 Ω – 1 kΩ
Capacitor, mica	1%	1%	0	±60	1 pF – 0.02 μF
Capacitor, poly-styrene	2%	2%	0	−150 ±60	20 pF – 0.1 μF
Capacitor, npo ceramic	5%	5%	0	0 ±30	10 pF – 0.01 μF

11.3   SAMPLED SIGNALS

Generally switched capacitor circuits sample a continuous analog input signal at a constant high frequency. As capacitive sensors usually use a high frequency clock to excite the sense electrodes and the sensor output is a clock-frequency variable amplitude signal, when using switched capacitor circuits to demodulate capacitive sensors, the sensor clock is a natural choice for the switched capacitor clock. The sampling process is shown in the time domain (Figure 11.1).

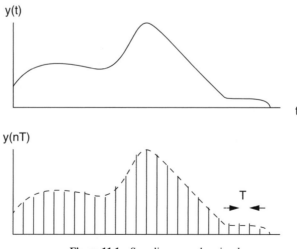

Figure 11.1   Sampling an analog signal

If the sampling frequency f_S (equal to $1/T$) is considerably higher than the maximum frequency of the analog input signal $y(t)$, say 4–10 times higher, the sampling process can be simply modeled as a delay of $T/2$ seconds without introducing serious errors. If, however, the input analog signal has frequency components which are higher than $f_S/2$, the signal cannot be completely recovered, and "aliasing" components are produced which may cause inaccuracies. This is the well-known sampling theorem, which is discussed in any book on digital signal processing [Bellanger, 1984, or Lynn et al., 1989]. Aliasing can be avoided by using a lowpass filter or a bandpass filter to remove high frequency input signals. These filters are also useful to remove unwanted interfering signals such as clocks which couple from other circuits in a system. Without attention, these clocks can alias down to low frequency signals which may be demodulated to a slowly varying signal offset.

If the sampling frequency is high compared to the maximum input frequency, classical AC analysis techniques will produce approximately correct results. If not, z-transform analysis should be used. The z transform is

$$Y(z) = \sum_{n=0}^{\infty} y(n)z^{-n} \qquad 11.1$$

A sampled input signal $y(n)$ is converted from an equation in time to an equation in the complex variable z.

The z transform is similar to the Fourier transform

$$S(f) = \sum_{n=-\infty}^{\infty} y(n)e^{-j2\pi fnT} \qquad 11.2$$

If a z transform analysis of a system is available, the frequency response of the system to a periodic waveform can be determined by replacing z by $e^{j2\pi fT}$.

The z transform is an essential analysis tool for digital signal processing, and the texts cited above or other DSP texts can be consulted for more information.

11.4   FILTERS

Switched capacitor filters substitute a switched capacitor

for a resistor. The input voltage, Vin, is sampled onto a small holding capacitor C during half of the square wave clock ω_c and discharged into a load impedance during the second half of the clock. The current *Iout* will be proportional to $CVin \times \omega_c$, and the circuit behaves like a resistor of value $1/C$. Note that the clock waveform does not need to be square; some designers use a highly asymmetric rectangular clock waveform to simplify output reconstruction filters.

MOS transistors can substitute for the switch

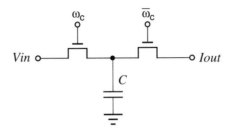

Propagation delay is added in the drive circuits so that ω_c and $\overline{\omega}_c$ have a slight underlap to protect against both transistors being momentarily on at the same time, causing a large current spike.

11.4.1 Lowpass filter, *RC*

A simple lowpass filter section (Figure 11.2)

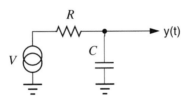

Figure 11.2   *RC* lowpass filter

has the familiar exponential response vs. time, with a *V*-volt step input

$$y(t) = V\left(1 - e^{-\frac{t}{RC}}\right) \qquad\qquad 11.3$$

$$t_1 = \frac{1}{RC}$$

and its frequency response, with a V-volt frequency sweep input, is

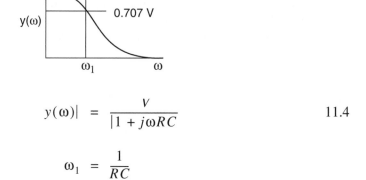

$$y(\omega)| = \frac{V}{|1 + j\omega RC}$$ 11.4

$$\omega_1 = \frac{1}{RC}$$

11.4.2 Lowpass filter, switched capacitor

A switched capacitor equivalent of the RC lowpass filter is shown in Figure 11.3.

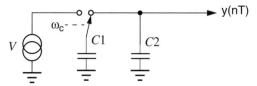

Figure 11.3   Switched capacitor lowpass filter

It has a similar exponential response vs. time

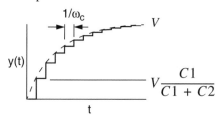

and a similar frequency response at low frequencies [Allen et al., p. 57]

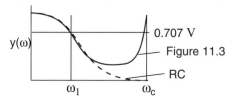

$$|y(\omega)| = \frac{V}{\left[1 + 2\dfrac{C2}{C1}\left(1 + \dfrac{C2}{C1}\right)(1 - \cos\omega T)\right]^{1/2}}$$ 11.5

$$\omega_1 = \frac{1}{T\left(1 + \dfrac{C2}{C1}\right)}$$

$$\omega_c = \frac{1}{T}$$

11.6

The switched capacitor lowpass filter can be designed to match the frequency response of the RC lowpass filter. Assume an RC filter with $\omega_1 = 20$ rad/s. First, a sampling frequency ω_c is chosen to be much larger than ω_1, say 5–25 times larger. We will use 500 rad/s. Then the values $C1$ and $C2$ are chosen so that

$$\frac{C2}{C1} = \frac{\omega_c}{\omega_1} - 1$$

which calculates as 3.98. The absolute value of $C1$ and $C2$ can be any convenient number for the chosen fabrication process. The accuracy of this method was verified by numerical methods and plotted in Figure 11.4.

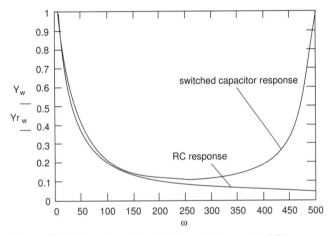

Figure 11.4   Comparison of switched capacitor lowpass to RC lowpass

This shows the expected good match at low frequencies and the poor performance when the input frequency approaches the sampling frequency. In most circuits, of course, an input lowpass (antialias) filter will be used and will reject frequencies above about 250 rad/s.

11.5   INTEGRATOR

Another circuit which works well with switched capacitor implementation is the integrator shown in Figure 11.5.

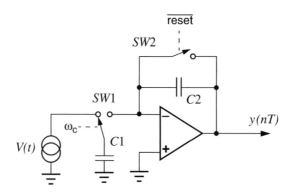

Figure 11.5   Switched capacitor integrator

Switch *SW2* is closed briefly at the beginning of the integration to discharge *C2*, then opened, and with each cycle of ω_c a charge packet equal to VC_1 is dumped into the inverting input of the amplifier. With a low-input-current amplifier, this charge is 100% transferred to *C2*. The voltage on *C2* is equal to the output voltage, and is

$$V_{C2} = \frac{Q}{C2} = \sum_{n=0}^{\infty} \frac{V_n \cdot C1}{C2} \qquad 11.7$$

V_n is the value of *V* at each sample time $1/\omega_c$. With a constant ω_c this circuit implements a sampled-data integrator with a gain of *C1/C2*.

11.6   INSTRUMENTATION AMPLIFIER

Switched capacitor circuits make a very effective instrumentation amplifier for converting differential signals with added DC into a single-ended signal [Allen et al., p. 81] (Figure 11.6).

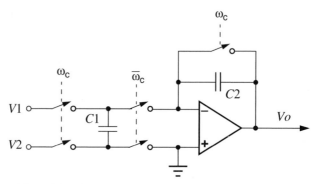

Figure 11.6   Switched capacitor instrumentation amplifier

During the first half-cycle of clock, the difference between input voltage $V1$ and $V2$ is impressed on $C1$ and $C2$ is zeroed. During the second half cycle, $C1$ is discharged into $C2$ which assumes a voltage

$$V_o = \frac{C1}{C2}(V1 - V2) \qquad 11.8$$

with a half-clock-cycle delay. V_o alternates with half-clock segments of 0 V, so a sample and hold circuit may be needed to remove the clock-frequency pulses. An advantage of this circuit over traditional discrete designs is that the common mode rejection is better; an integrated switched capacitor instrumentation amplifier is available from Linear Technology in a 16-pin package, the LTC1043, with a common mode rejection ratio of 120 dB and a maximum clock rate of 5 MHz.

Differential integrator

In the circuit above, if capacitor $C2$ is much larger than $C1$ and the switch which zeros $C2$ is turned on less frequently, the same circuit becomes a differential integrator. $C2$ is zeroed at the beginning of a measurement cycle, then $C1$ is switched many times to accumulate input voltage samples in $C2$. No output sample-and-hold function is needed with this use.

11.7   DEMODULATOR

The synchronous demodulator often used to demodulate a capacitive sensor which was shown in Chapter 4 (Figure 4.5) is repeated here (Figure 11.7). This demodulator can be integrated simply, using a switched capacitor circuit (Figure 11.8).

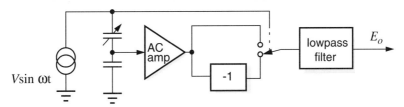

Figure 11.7   Synchronous demodulator

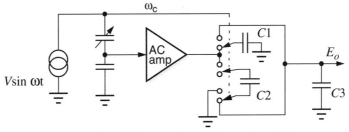

Figure 11.8   Switched capacitor synchronous demodulator

The gain is proportional to $C1/C3$ for positive half cycles of ω_c and proportional to $C2/C3$ for negative half cycles. Usually $C1 = C2$ and $C3 > C2$ to produce a lowpass response and the lowpass frequency is determined by $C3$ as above (Figure 11.3). E_o is a discontinuous-time baseband signal.

11.8   SWITCHED CAPACITOR CIRCUIT COMPONENTS

11.8.1   Switch

The lowpass switched capacitor circuit (Figure 11.3) works with an approximately square-wave clock, ω_c. When the switch is in the left position a charge $Q = C1 \times V$ is accumulated in $C1$. The switch then transfers most of the charge to the (usually) larger capacitor, $C2$. The first step size is $VC1/(C1+C2)$, and as $C2$ charges toward V, proportionately less charge is transferred to produce a decaying exponential.

As in all switched capacitor circuits, the accuracy of the switch is critical. ON resistance, Rds_{ON}, and injected charge, Q_I, are the most important parameters for integrated switches; with discrete MOS switches or packaged integrated switch arrays (CD4066 or DG401, as examples), the leakage current must also be considered.

The switch can be fabricated with small-geometry n-channel MOSFETs for integrated designs. Table 11.3 lists the circuit parameters when 5 μm silicon rules are used.

Table 11.3 Component parameters for switched capacitor lowpass

Component	Parameters	Performance
n-FET, silicon, 5 μm rules	$Rds_{ON} = 2\text{–}10\ k\Omega$ $Q_I = 0.05\ pC$ $I_S\ (off) = 0.2\ pA$	$dv/dt = 0.02$ V/s in hold circuit
FET switch, discrete, D444	$Rds_{ON} = 85\ \Omega$ $Q_I = 5\ pC$ $I_S\ (off) = 10\ pA$	$dv/dt = 1$ V/s in hold circuit
$C1$	1 pF	total charge = 5 pC @ 5 V
$C2$	10 pF	charge time @ 2000 Ω = 60 ns

A small-geometry integrated FET works well with the small capacitance values available on silicon. The ON resistance at 2 kΩ is low enough to charge the large 10 pF capacitor in less than 60 ns. Q_I, the injected charge, is the small packet of charge which is transferred to the negative supply with each switch transition and should be considerably less than the charge stored on $C1$ for good accuracy.

With the silicon FET the injected charge is 1% of $C1$'s charge. The discrete FET injected charge is equal to $C1$'s charge, so this FET would be hopelessly inaccurate when used with this small capacitor. With a discrete capacitor in the 100–1000 pF range, however, the discrete switch is capable of good results.

11.8.2 Operational Amplifiers

The operational amplifier for silicon implementation may be a MOS type with the following equivalent circuit [Allen et al., pp. 691, 695] (Figure 11.9).

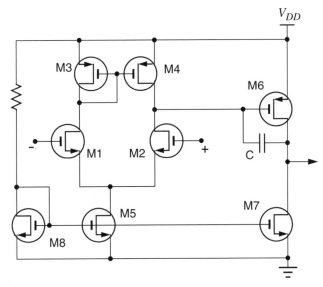

Figure 11.9   MOS amplifier equivalent circuit

The amplifier uses the standard differential-input amplifier, $M1$-$M2$-$M5$, which contributes high differential gain and low common mode gain. Its load is the high impedance Wilson current source $M3$ and $M4$, and $M6$ buffers and further amplifies the signal. The capacitor C, sometimes with a series resistor, shapes the amplifier response to a –6 dB/octave rolloff for stability, and the value of C is adjusted to the closed loop gain of the amplifier for best frequency response. The voltage gain A, if the input pair $M1$ and $M2$ are identical, is

$$A = \frac{g_{m2}}{g_{ds2} + g_{ds4}} \cdot \frac{g_{m6}}{g_{ds6} + g_d}$$

$$g_m = \sqrt{2\mu C_{ox}(W/L)I_D} \qquad\qquad 11.9$$

$$g_{ds} = \lambda I_D$$

The transconductance g_m is a function of μ, the surface mobility of electrons or holes, C_{ox}, the gate oxide capacitance per unit area, and the FET's channel width W and length L. The channel conductance g_{ds} (the reciprocal of channel resistance) depends on λ, the channel length modulation factor determined by the particular CMOS process used, and the drain current I_D. A SPICE analysis of this amplifier shows the characteristics listed in Table 11.4.

Table 11.4 MOS operational amplifier characteristics

Gain, Av	5000
Rout	500 kΩ
Unity gain bandwidth	1 MHz
Phase margin, no load	75°
Input offset voltage	10 mV
Power consumption	2.5 mW
Power supply rejection ratio	82 dB
Common mode rejection ratio	80 dB
Slew rate with $C_L = 10$ pF	10 V/μs

To achieve this performance, the width-to-length ratios of the FETs are customized, from a small geometry 10/10 input FET to larger 100/10 (100 μm width, 10 μm length) output FETs. These characteristics are similar to packaged operational amplifiers, except the gain is considerably lower and the output resistance is considerably higher for the MOS implementation, but neither of these compromises causes too much trouble with integrated switched capacitor circuits. DC gain is not critical in circuits where the signal is modulated on a 20 kHz carrier, and the gain at 20 kHz is limited to 50× by the 6 dB/octave compensation and the unity gain bandwidth. The high value of output resistance for the MOS circuit, 500 kΩ, would be awkward in a packaged amplifier, but the amplifier is typically driving very high impedances if it is used on-chip; an amplifier which drives signals off-chip would need an additional buffer for lower output impedance.

Bi-MOSFET amplifier

A low noise, low input capacitance integrated amplifier can be built with bipolar-MOS processes, using MOS input stages and bipolar outputs. This combination gives high input impedance, near-zero current noise, and low output impedance for driving low impedance loads, capacitive loads, or driving off-chip. A bi-MOS amplifier designed for biological probes [Takahashi et al., 1994] was optimized for high input impedance and low input capacitance; it is described in this paragraph.

MOSFETs are noisier than bipolar transistors or JFETs, but input-referred noise voltages of 15 nV/$\sqrt{\text{Hz}}$ can be achieved with p-channel FETs and current noise is very low. Current noise is normally a problem only when a MOSFET input needs to be brought out to a bonding pad; the usual diode protection for electrostatic discharge can contribute much higher current noise than the amplifier. The Bi-MOS amplifier uses a p-channel MOSFET input differential amplifier, a p-MOS second stage, and a Darlington bipolar output stage. The design challenge is to achieve low noise with low input capacitance.

For the BiFET amplifier input circuit transistors, transconductance and midband noise decrease by W/L while capacitance is minimized by reducing WL, so a high W/L ratio of 50 is chosen. A capacitance of 6 pF and a spot noise of 38 nV/$\sqrt{\text{Hz}}$ is achieved well into the $1/f$ region at 280 Hz. Additional information on FET noise is found in Section 12.1.2.

11.9   ACCURACY

11.9.1   Capacitor ratios

Switched capacitor circuits have two very favorable characteristics for silicon implementations. The gain, for most circuits, is proportional to the ratio of capacitors. Capacitor ratios are the most stable parameter available on an integrated circuit, as seen in Table 11.1, and rival the stability of RC products available in precision discrete components.

11.9.2   Amplifier accuracy

Amplifier input offset voltage accuracy can be excellent, contributing at worst a few mV of DC offset. Most sensitive capacitive sensor circuits amplify the signal while it is modulated on the carrier frequency, so the DC offset is unimportant except at the final demodulator stage. For integrated amplifiers as above, the relatively low gain compared to packaged ICs may contribute to small changes in gage factor and imperfect cancellation of stray capacitance for guard circuits.

11.9.3   Parasitics

The MOS capacitor uses SiO_2, glass, as a dielectric. This is one of the most stable and least lossy dielectrics available. Parasitic capacitors are formed, however, with the typical MOS capacitor construction.

The silicon oxide isolation and passivation layer forms the dielectric for the MOS capacitor. Isolation from the substrate is done with the usual n^+ diffusion, forming a reverse biased diode and also acting as a reasonably low-resistivity bottom plate. The top plate is the aluminum interconnect metallization. Both top and bottom plates will add a small parasitic capacitance, C_T and C_B, to the capacitor C (Figure 11.10).

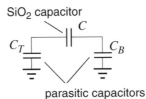

Figure 11.10   Parasitic MOS capacitances

The effect of these capacitors is usually to just slightly increase another circuit capacitor value. Suppose, for example, C_T and C_B are considered in the demodulation circuit above (Figure 11.8). Considering $C1$, C_B is shorted and C_T simply adds to $C1$. For $C2$, C_T is charged by the amplifier but does not contribute to the output, while C_B adds to $C3$. For $C3$, C_T adds its capacitance and C_B is shorted and does not contribute to the output. If the MOS capacitor value is adjusted downward by a few percent to compensate for the parasitics; no other circuit effects will be seen. C_B's value ranges from 5–20% of the total [Allen et al., p. 397], while C_T is between 0.1 and 1%.

11.9.4 Charge injection

A small packet of charge is injected into a MOS switch with each clock edge. It is caused by small voltage-dependent parasitic capacitances from gate to source and drain electrodes, and varies from 5 pC for discrete circuits to 0.005–0.05 pC for small-geometry IC switches, as shown in Table 11.3. Uncompensated charge injection can be a serious source of error; for ICs, sample and hold errors in the range of 5–50 mV are typical. The amount of charge injection usually increases as switching speed increases, and also varies with the DC bias point. Several techniques can be used to minimize this charge. A simulation of charge-injection errors in a D/A converter using a variety of circuit techniques to minimize and compensate for injection errors [Willingham et al., 1990] shows an improvement of about 500×. Actual circuits will not achieve this level of performance, but the simulation shows that performance is limited by component tolerance and not by systematic design flaws.

Dummy switches

A dummy switch can be connected in a way that injects an equal and opposite quantity of charge. A simple circuit with C_S representing the signal input port and C_L the output shows the effects of charge injection. An unwanted voltage step is seen at the output (Figure 11.11).

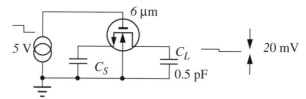

Figure 11.11   Charge injection

With typical 3 μm self-aligned p-well n-channel FETs, the injected charge at a DC level of 2.5 V is 0.01 pC which causes a voltage step (from $Q = CV$) of 20 mV in a 0.5 pF capacitor. If an inverted clock is used to drive a half-sized dummy transistor (Figure 11.12)

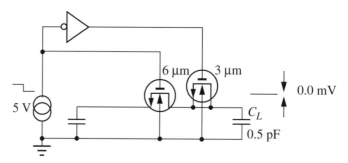

Figure 11.12   Charge injection, dummy switch

the injected charge can, in theory, be exactly nulled. A 3 µm wide transistor with an inverted drive for charge injection compensation is used in parallel with the 6 µm working switch in Figure 11.11. With the normal symmetric transistor, the gate-to-drain capacitance is the same as the gate-to-source capacitance, and only the gate-to-drain capacitance of the working switch contributes to charge injection if the source C_S is respectably low impedance, hence the 2:1 ratio. But two effects mitigate the success of this effort: if clock rise and fall times are slow or if the logic inverter has finite delay, half-size transistors (or, more accurately, half width-to-length-ratio transistors) are not optimal; the size can be adjusted to be somewhat smaller for improved performance. Also, as the parasitic capacitors are voltage-dependent, the exact ratio also depends on the threshold voltage [Eichenberger, 1990] and should also be adjusted to the signal input voltage. With a large input signal swing, all factors cannot be accommodated at once.

Subtraction

Another method [Martin, 1982] can null charge injection more accurately by subtraction in circuits where an operational amplifier is used. The integrator in Figure 11.5 is redrawn with individual MOS switches in Figure 11.13.

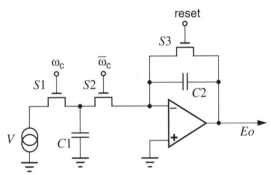

Figure 11.13   Integrator

Transistor $S1$ will pull some charge from $C1$ when it is shut off, and transistor $S2$ will inject charge into $C1$ and $C2$. These effects can be compensated with a similar circuit feeding the noninverting input, as shown in Figure 11.14 on page 193.

11.10   BALANCED DIFFERENTIAL-INPUT DEMODULATOR

A balanced circuit which features:

- Linearization of 1/spacing dependence
- Reference capacitors to compensate environmental effects
- Differential processing to minimize charge injection

has been described [Schnatz et al., 1992] for use with a pressure transducer, but it demonstrates excellent techniques for accurate switched-capacitor demodulation for any type of sensor. The transducer described has a measured nonlinearity of less than 0.4% from pressure input to voltage output. Its block diagram is shown in Figure 11.15.

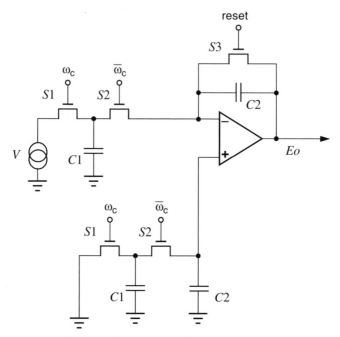

Figure 11.14 Integrator with charge cancellation

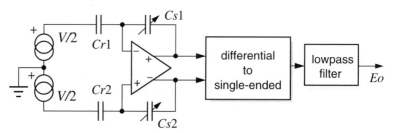

Figure 11.15 Balanced demodulator block diagram

The sense capacitor is divided into two equal capacitors, $Cs1$ and $Cs2$, and a reference capacitor Cr in close physical proximity is divided into two equal capacitors, $Cr1$ and $Cr2$. The output voltage is

$$E_o = V \cdot \frac{Cr}{Cs}$$

which is linear for spacing-variable capacitors; if Cs is an area-variable capacitor, its position can be interchanged with Cr. The circuits are implemented using switched capacitor techniques (Figure 11.16).

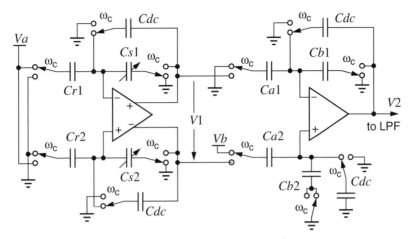

Figure 11.16   Balanced demodulator circuit

All switches are shown in the $\omega_c = 1$ position. The first stage produces a differential DC output voltage with amplitude proportional to the plate spacing. With $\omega_c = 1$, $Cr1$ is charged to Va, $Cs1$ is discharged, and Cdc holds the previous output voltage sample; $Cr2$, etc., do the same with negative polarity. When ω_c switches to 0, $Cr1$'s charge is transferred to $Cs1$ and Cdc is charged to the output voltage. The exact value of Cdc is unimportant, except it must be large enough to handle worst case leakage currents without excessive droop and small enough to charge fully in one cycle of clock. The amplifier output takes the value $V1 = -2VaCr/Cs$; uncompensated voltage drops, interfering signals, charge injection, and leakage current effects appear as a common mode voltage. Appropriate small delays must be added to the ω_c signals to avoid simultaneous conduction.

The following stage is the switched capacitor version of the standard four-resistor instrumentation amplifier; its common mode rejection is equal to the matching of the capacitor ratios and can be 40 dB with careful processing. This stage cancels the unwanted common mode voltage component of $V1$ and adds an adjustable DC offset, Vb, if needed. $Ca1$ and $Ca2$ are nominally equal, as are $Cb1$ and $Cb2$; the dummy switch connected to $Cb2$ compensates for the charge injected by the switch connected to $Cb1$. If the amplifier is to produce an output centered around a displaced value, Vb is adjusted appropriately; otherwise Vb can be zero. This stage can also add gain as needed to establish the gage factor; its gain is Ca/Cb. The overall transfer function is

$$V2 \; = \; \frac{Ca}{Cb}\left(2Va\frac{Cr}{Cs} - Vb \right)$$
11.10

11.11   ALTERNATE BALANCED DEMODULATOR

Using the instrumentation amplifier above (Figure 11.6) and simplifying the input stage, a simplified balanced demodulator is as shown in Figure 11.17.

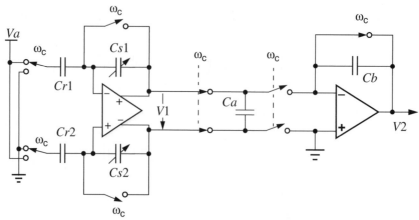

Figure 11.17   Alternate balanced demodulator

Again, all switches are shown in the $\omega_c = 1$ position. The first stage output $V1$ is an amplitude-modulated square wave rather than the discrete-time analog signal shown in Figure 11.16. The variable capacitors $Cs1$ and $Cs2$ are shorted with $\omega_c = 0$, and with $\omega_c = 1$ the fixed charge impressed upon the reference capacitors $Cr1$ and $Cr2$ is transferred to $Cs1$ and $Cs2$. $V1$ is picked up by Ca when $\omega_c = 1$ and its charge is transferred to Cb with $\omega_c = 0$ to complete the balanced-to-single-ended conversion. The common-mode rejection of this circuit is excellent as it is not dependent on component matching. The $Ca - Cb$ circuit will need charge injection compensation for maximum accuracy as shown in Figure 11.12, and the square wave output can be converted to DC with a sample and hold, a low-pass filter, or the same Cdc switching shown in Figure 11.16. This circuit will need more power to bias the operational amplifiers, as higher slew rate is needed than the previous circuit to handle the square wave output signals. The overall circuit gain is

$$V2 = -\frac{Ca}{Cb}\left(2Va\frac{Cr}{Cs}\right) \qquad 11.11$$

12

Noise and stability

Capacitive sensor performance is limited by circuit noise from various sources, by mechanical stability, and by environmental factors. If these limits are understood and handled correctly, capacitive sensor designs can have exceptionally stable and low noise performance.

Circuit and component noise was briefly discussed in Section 10.4, and the level of detail in that chapter should be adequate for almost all applications, but for the small percentage of design jobs which "push the envelope," more information is presented in this chapter and the limiting sensitivity of capacitive sensors is analyzed.

12.1   NOISE TYPES

Five different types of noise can be found in sensor preamplifiers. All five types are present to some degree in typical circuits, but careful design can minimize the noise contribution of all except thermal noise.

Thermal noise

Thermal noise is broadband noise, having an energy which is a function only of the measurement bandwidth and not the frequency of measurement. It is generated by resistive components only; reactances do not generate thermal noise. It is caused by the thermally excited movement of molecules, and disappears at a temperature of absolute zero. Thermal noise generated by resistors follows a simple equation; thermal noise generated by semiconductors can also be predicted by understanding more complicated semiconductor processes. To simplify analysis, all semiconductor noise effects are given as equivalent input voltage and current noise sources.

Shot noise

Shot noise is simply due to the fact that the current through a semiconductor junction is quantized by the electron charge, and electron arrival time has a Gaussian distribution. Shot noise is generated any time that electron flow is gated by a valve which lets through one charge carrier at a time, such as the emission of an electron by a vacuum-tube cathode or a minority carrier in a transistor falling through a potential barrier. In our circuits, shot noise is caused by diodes and transistors. Conductors and resistors do not generate shot noise, as a fraction of a charge carrier can be transmitted by these devices. Shot noise I_s is characterized by this equation

$$I_s = \sqrt{2qI\Delta f}$$ 12.1

$q = 1.60 \times 10^{-19}$ C, electron charge

Δf = measurement bandwidth, Hz

I = DC current through junction

1/f noise

Both resistors and semiconductors have a low frequency noise component which varies as approximately $1/f$. This noise is produced by a variety of process-dependent mechanisms and is characterized empirically.

Popcorn noise

Popcorn noise is found only in semiconductors, both bipolar and FET, and produces small, abrupt steps of output voltage at a low and random repetition rate. Popcorn noise was named by early users of operational amplifiers for audio amplifiers. They noticed a low level noise in the audio output which sounded like popcorn popping. Popcorn noise is associated with defects in the semiconductor crystal [Fish, p. 88] and has improved in recent years with increasing purity of semiconductor materials. It is a rapid change in a semiconductor bias current between two extremes (Figure 12.1). The level can be 5–20 times the Gaussian noise level. Popcorn noise will rarely be a problem with newer amplifier types.

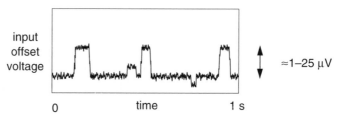

Figure 12.1   Popcorn noise

Crosstalk

Circuit crosstalk is the inadvertent coupling of voltages from other circuits to the sensor amplifier. Digital logic signals are particularly unfriendly, as crosstalk from a number of digital signals can look on an oscilloscope like random noise, and the signals can fall in the bandwidth of the capacitive sensor amplifier. Crosstalk is much easier to handle for high impedance capacitive sensor signals than for low impedance inductive sensor types, as the electric field is almost perfectly attenuated by conductive shielding while low impedance shielding usually must have magnetic field attenuation also.

12.1.1 Resistor noise

Resistors, which are carrying no current, produce thermal (Johnson) noise. DC voltage across a resistor causes an additional low frequency noise component with an approximately $1/f$ spectrum, called excess noise [Fish, 1994, pp. 184–190; Motchenbacher, 1993, pp. 289–296]. The excess noise changes considerably with the resistor size and construction method. The rms noise from these effects is

Thermal noise

$$e_t = \sqrt{4kTR(f_2 - f_1)} \quad \text{V rms} \qquad \qquad 12.2$$

where

$k = 1.38 \times 10^{-23}$, Boltzmann's constant, J/K

$T = 273 + °C$, absolute temperature, K ($4kT$ at 25°C $= 1.64 \times 10^{-20}$)

$f_1 =$ lower 3 dB point, Hz

$f_2 =$ upper 3 dB point, Hz

$R =$ resistance, Ω

Excess noise

$$e_x = C_F Vr \sqrt{\log(f_2/f_1)} \quad \text{V rms} \qquad \qquad 12.3$$

where

$Vr =$ voltage across resistor

$C_F =$ excess noise factor, empirical

These equations assume a rectangular bandpass filter is used with a lower 3 dB frequency of f_1 and an upper 3 dB frequency of f_2.

The resistor excess noise factor C_F varies from 0.01–1 μV $\times V_{DC}^{-1} \times$ (freq decade)$^{-2}$ [Fish, 1994, p. 88]. For a resistor with 1 V drop and 0.1 μV of excess noise in a decade of measurement bandwidth, C_F is 1. C_F is also called NI, or noise index, which is usually expressed in dB.

As an example, assuming $C_F = 10^{-7}$, temperature is 25°C, resistance is 1 MΩ, $Vr = 1$ V, and the 3 dB frequencies of the bandpass are 10 kHz and 20 kHz, the resistor noise is

$$e_t = 12.8 \times 10^{-6} \text{ thermal noise}$$

$$e_x = 3.01 \times 10^{-8} \text{ excess noise}$$

Shot noise

Shot noise will not be seen in a circuit with just a resistor, but if a diode or semiconductor junction is in series, the Gaussian shot noise component will be added, as current transfer through a potential well is quantized by electron charge. For a large value resistor the shot noise contribution is insignificant, less than 60×10^{-12} for the 1 MΩ resistor above.

Excess noise is Gaussian and has a noise spectrum which falls off with frequency

$$e_x = \frac{K_f}{f^\alpha} \qquad\qquad 12.4$$

where α is usually in the range of 0.8–1.4 [Fish, 1994, p. 85] for semiconductors and 0.8–1.2 for resistors. We use the median value of 1. With $\alpha = 1$, the rms value of the excess noise measured in the bandwidth between 1 Hz and 10 Hz is the same as the rms excess noise measured in a bandwidth of 10 Hz to 100 Hz. Since excess noise is proportional to the number of decades of measurement frequency bandwidth and is independent of where in the spectrum the bandwidth is measured, it is specified as microvolts per volt in one decade. If the circuit bandwidth is not one decade, the noise is scaled as

$$e_t' = e_t \sqrt{\frac{\log(f_2/f_1)}{\log(10)}} \qquad\qquad 12.5$$

Philips [*Precision MELF SMD Resistors*, 1992, p. 106] shows excess noise curves for two sizes of surface mount resistors. The small size, 0805, has an excess noise of 0.5 μV/V in a decade at the 100 kΩ value; the larger 1406 size shows a value of 0.15 μV/V for the same conditions. This reference also shows an increase in excess noise with resistance. Excess noise increases as $\sqrt{R}$ for values under 500 kΩ and as about $R^{0.75}$ for values above 500 kΩ. The noise of a 1406 size 5 MΩ precision resistor in one decade of frequency, from this reference, is 2 μV/V. For this resistor,

- Thermal noise = 28 μV
- Excess noise = 10 μV (1 kHz – 10 kHz bandwidth, 5 V DC)

The excess noise performance of all construction methods worsens at high resistance values and small sizes. Also, the shape of the excess-noise-vs.-resistance graph is unpredictable; at 1 kΩ a carbon film resistor [Motchenbacher, 1993, p. 295] has an excess noise of –37 dB, or –97 dB below the applied voltage (add 60 dB as the excess noise C_F is μV/V), while a 1/2 W carbon composition resistor has –33 dB excess noise. At 10 MΩ, however, these figures worsen and reverse, with the carbon comp resistor at –10 dB and the carbon film resistor at +5 dB. The lesson for capacitive sensor preamps is to *design the input circuit so high-value resistors have no voltage drop*. This is not difficult if low bias current op amps or FETs are used.

12.1.2 FET noise

Transistor noise has three components:

- Excess noise is a problem at frequencies of less than 10 Hz–10 kHz. It follows an approximate $1/f$ characteristic.

- Thermal noise is independent of frequency, and is defined as the noise voltage found in 1 Hz of bandwidth. It increases as the square root of bandwidth, and is caused by the thermal noise of the FET channel resistance.
- Shot noise is independent of frequency like thermal noise.

The total noise increases at high frequencies, in excess of 100 kHz–10 MHz, primarily due to current noise increase as thermal noise in the channel resistance is coupled back to the gate by Miller capacitance (Figure 12.2).

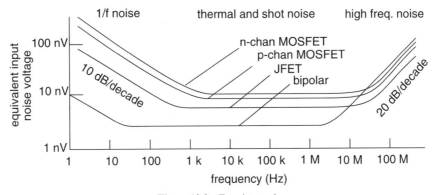

Figure 12.2   Transistor noise

These curves are very approximate; individual transistor processes can result in 10× more noise than shown. GaAs FETs are not shown, but are not useful for capacitive sensors as they have a very high 1/*f* noise corner, in the range of 1 MHz–10 MHz.

Current noise is 3–5 orders of magnitude higher in bipolar transistors than in JFETs or MOSFETs, rendering them unacceptable for our high-impedance sensing problems; they are shown for comparison only.

Transistor noise can also be separated into voltage and current noise, and these components behave differently vs. frequency for FET transistors [Motchenbacher, 1993] (Figure 12.3).

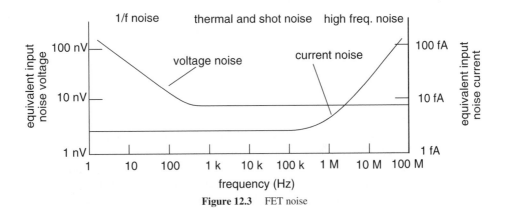

Figure 12.3   FET noise

Bipolar transistors have a total-noise-vs.-frequency characteristic similar to FETs, but the components are interchanged. For bipolars [Motchenbacher, 1993, p.126] the noise voltage shows no $1/f$ noise, except perhaps at high currents and low frequencies, but it increases for frequencies above the transistor f_T, and the noise current shows $1/f$ behavior as well as a high frequency rise. For bipolars and FETs, the $1/f$ portion of the curve above graphs noise *power* changing as $1/f$, so noise voltage changes as roughly $1/\sqrt{f}$ or 10 dB/decade, although process variations may produce a much lumpier curve. The high-frequency slopes are all at the 20 dB/decade rate of a one-pole response.

Table 12.1 Discrete transistor noise measurements

	Bipolar	MOSFET	JFET	JFET op amp
E_n, nV/$\sqrt{\text{Hz}}$, 10 kHz	0.9–3.6	5–6	0.8–6	10–43
I_n, pA/$\sqrt{\text{Hz}}$, 10 kHz	1–4	0.0002–0.0007	0.0035–0.035	0.0001
$1/f$ corner of E_n	none	10 kHz–500 kHz	1 kHz–10 kHz	
High freq. corner of E_n	100 kHz–10 MHz	none	none	
$1/f$ corner of I_n	160–3000 Hz	none	none	
High freq. corner of I_n	none	10–200 kHz	10–500 kHz	
Tested types include	2N4250, 2N4403, 2N4058, 2N4124, 2N4935, 2N4044	2N3631 HRN1030	2N2609, 2N3460,2N3684, 2N5116,2N5266, 2N5394, C413N, UC410	see Table 10.2
Lowest noise type	2N4403	2N3631	C413N	

A sample of low noise transistors was tested by Motchenbacher [1973, pp. 78, 106] and excerpted here in Table 12.1. For a discrete FET, the equivalent input-referred thermal-noise resistance can be estimated as 0.66 times the reciprocal of transconductance, $0.66/g_m$. As an example, a FET transistor with a g_m of 1 S (1 mho) is equivalent to a noiseless FET with a thermal-noise-generating 1 kΩ resistor in series with the output. The input-referred $1/f$ noise of a MOSFET, E_n, is proportional [Takahashi, p. 390] to

$$E_n = \sqrt{\frac{d^2}{\varepsilon^2 W L f}} = \frac{\sqrt{WL}}{C} \qquad 12.6$$

E_n is the rms spot noise at frequency f, d and ε are the thickness and the dielectric constant of the gate insulation, W and L are the width and length of the channel, and C is the gate capacitance. As the WL product increases, noise decreases. Unfortunately, many applications cannot take advantage of a large area because the input capacitance also increases directly with WL. Thermal noise decreases directly as the conductance g_m increases, and as g_m is directly related to W/L, thermal noise will decrease as $W^{3/2}L^{-1/2}$.

A more complete description of MOSFET input noise [Motchenbacher, 1993, p.150; Sze, p. 496] is

$$E_{ni}^2 = \frac{0.66}{g_m} \cdot 4kT + \frac{K}{C_{ox}WLf} \qquad 12.7$$

The first term is thermal noise and the second term is $1/f$ noise.

$$I_{ng}^2 = 2qI_{DC} \qquad 12.8$$

E_{ni} = input-referred voltage noise, rms, per $\sqrt{Hz}$

0.66 = empirically determined noise constant, ranging from 0.5–1

g_m = transconductance, $1/\Omega$ or S (Siemens)

kT = Boltzmann constant × temperature, K; 4.14×10^{-21} at 27 °C

K = flicker noise coefficient, in the range of 10^{-30}

C_{ox} = capacitance per unit area of oxide, farads per μm^2

I_{ng} = gate noise current, rms per $\sqrt{Hz}$

q = electron charge, 1.602×10^{-19} C

I_{DC} = gate leakage current

A typical low-noise MOSFET with gate leakage of 100 fA, $C_{ox} = 0.7$ fF/μm^2, $W/L = 50$, and a channel length L of 3.2 μm was calculated [Motchenbacher, 1993, p. 151] as

$$E_{ni} = 8.19 \text{ nV}/\sqrt{Hz}$$

$$I_{ng} = 0.18 \text{ fA}/\sqrt{Hz}$$

These specifications are typical of n-channel JFET low noise amplifiers (Table 12.2).

Table 12.2 N-channel JFET low noise amplifier transistors

Type	2N5558	2N5105
Process	NJ16	NJ26AL
Gate leakage, typ	10 pA	10 pA
Transconductance g_m	0.0022 S	0.009 S
Input capacitance	3 pF	5 pF
Feedback cap.	1 pF	1.5 pF
Noise voltage, En	6 nV/$\sqrt{Hz}$	1 nV/$\sqrt{Hz}$
$1/g_m$	454 Ω	111 Ω
Equiv. noise resistor	375 Ω	62.5 Ω

The equivalent noise resistor is calculated from *En* and *In* using eq. 12.2.

Low-Z amplifier

The noise equivalent circuit of the inverting or low-Z amplifier is shown in Figure 12.4. If the amplifier gain is high, the output noise voltage assuming uncorrelated noise sources and measured in a bandwidth Δf at a frequency ω higher than the $1/Rf\,Cf$ pole is

$$Eo^2 = En^2\left[1 + \frac{Cs + Cg}{Cf}\right]^2 + \frac{4kT\Delta f}{\omega^2 Cf^2 Rf} + \left[\frac{In}{\omega Cf}\right]^2 \qquad\qquad 12.9$$

Noise voltages are in volts $\sqrt{\text{Hz}}$ and noise currents are in amps $\sqrt{\text{Hz}}$ rms. The first term is the effect of the input voltage noise, the second is the contribution from the thermal noise of *Rf*, and the third is the effect of the input current noise. If the resistor *Rf* is large compared to the sensor impedance, the equation becomes

$$Eo^2 = En^2\left[1 + \frac{Cs + Cg}{Cf}\right]^2 + \left[\frac{In}{\omega Cf}\right]^2 \qquad\qquad 12.10$$

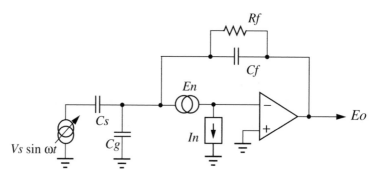

Figure 12.4   Inverting amplifier noise equivalent circuit

12.1.3 High-impedance amplifier

The noninverting or high-Z amplifier noise circuit is shown in Figure 12.5.

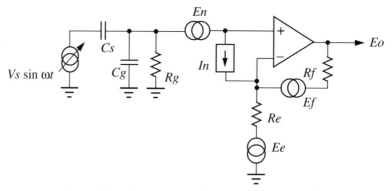

Figure 12.5   Noninverting amplifier noise equivalent circuit

Noiseless feedback resistors Re and Rf have been added to establish a gain $G = 1 + Rf/Re$, and each has a thermal noise generator in series. The output noise of this circuit, assuming the impedance of the parallel combination of Cs, Cg, and Rg is greater than Re and defining $Ct = Cs + Cg$, is

$$Eo^2 = Ef^2 + G^2 En^2 + \left(Ee\frac{Rf}{Re} \right)^2 + G^2 \frac{4kTRg \cdot \Delta f}{(1 + sRgCt)^2} + G^2 In^2 \frac{Rg^2}{(1 + sRgCt)^2}$$

Simplifying the circuit with a gain of 1, with $Rf = 0$ and $Re = $ infinity, and again assuming that Rg is large enough so that its noise contribution is negligible, we have

$$Eo^2 \cong En^2 + \left[\frac{In}{\omega(Cs + Cg)} \right]^2 \qquad \qquad 12.11$$

If Cf is set equal to Cs so that the gains of the two circuits are equal, and if $Cg = 0$, the high-Z amplifier has a 2× advantage in output signal-to-noise ratio. If, however, a shield is used with a capacitance Cg to the pickup electrode, and it is either grounded for the low-Z amplifier or used as the feedback capacitor for the high-Z amplifier, the output noise for the two circuits is identical. As shields are almost always used, changing from high-Z to low-Z does change gain and output level, but not signal-to-noise ratio.

The feedback amplifier noise performance is similar to the high-Z circuit.

12.2   LIMITING DISPLACEMENT

The just-measurable limiting displacement in a capacitive position sensor x_n is the displacement which produces a voltage change equal to the rms noise voltage in the bandwidth of the amplifier circuit. The just-measurable displacement in meters is defined as

$$x_n = \frac{Eno}{G} \qquad \qquad 12.12$$

where G is the responsivity of the sensor in volts per meter of displacement at the amplifier output and Eno is amplifier output noise.

12.2.1   Limiting displacement of three-plate micrometer

Jones [1973] defines the limits of performance of three-plate micrometers. With a low-Z amplifier, we have the circuit shown in Figure 12.6. With frequency high enough so that the crossover resistor En/In is much larger than the impedance of $C1 + Cd$, the capacitance change in $C1$, $\Delta C1$, which produces an output equal to the rms noise is

$$\Delta C1 = (C1 + Cd)\frac{En}{0.707 V} \qquad \qquad 12.13$$

with En evaluated in the bandwidth of BPF.

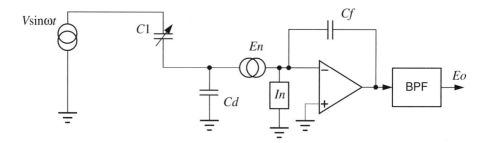

Figure 12.6 Micrometer circuit

With small sensor capacitance and low excitation frequency the current noise *In* may not be small enough to be neglected, and consideration should be given to increasing the excitation frequency. To detect small changes of absolute capacitance, increase excitation voltage *V*, use a low noise transistor, and reduce *C*1 until *C*1 is approximately equal to the stray capacitance *Cd* or until the current noise contribution becomes the dominant noise source.

12.2.2 Maximizing limiting displacement

If $C1 = Cd = 1$ pF and the excitation voltage is 100 V, with a 1 nV/$\sqrt{\text{Hz}}$ amplifier and a bandwidth of 1 Hz, the highly theoretical minimum detectable absolute capacitance is an exhilarating 2×10^{-22} F, 0.0003 aF. This represents a 2×10^{-10} change, so with a plate spacing of 1 mm a displacement of 2×10^{-13} m, 0.04 light wavelengths, is just detectable.

Coupling transformer

A largely theoretical option is to use a wideband input coupling transformer. Ideally, an amplifier with the lowest possible input noise *power* (*In*× *En*) should be selected and the sensor impedance matched to the amplifier's *En*/*In* using a transformer. The transformer turns ratio should be the square root of the ratio of the sensor impedance to *En*/*In*. As this amplifier selection criterion would result in very high impedance, several hundred MΩ, the transformer would need an inductance of at least 600 H at a 25 kHz carrier frequency along with very low parasitic capacitance. This is an unrealizable component. Adding a transformer to the drive to increase excitation voltage is preferable, at least until air breakdown happens.

Tuned transformer

A more realizable alternate is a tuned input transformer. This should be used with sine wave drive, or at least with a demodulator which accepts sine wave signals, and it optimally has a turns ratio selected as above. If the transformer inductance is tuned with a shunt capacitor at the signal frequency, any small value of inductance is resonated to a high impedance if the inductor's *Q* is sufficiently high. The tuned transformer circuit has two useful features, as it both reduces noise by bandwidth reduction by a factor proportional to *Q*, and it matches amplifier noise resistance. Jones [1973, p. 594] uses a tuned transformer and a 100 V rms sine wave drive signal to achieve these impressive performance numbers

- Transformer $Q100$
- Output resolution, 1 Hz bandwidth0.01 aF $(0.01 \cdot 10^{-18}$ F)
- Minimum detectable motion$5 \cdot 10^{-6}$ μm

Tuning the input with a Q of 100 can produce a circuit with an equivalent noise resistance of 100 Ω [Jones, p. 595]. The equivalent noise voltage of this resistance is, from eq. 10.9 above and using a bandwidth of 100 Hz, 12.9 nV, for a 170 dB dynamic range in a 5 V system. This performance is only a factor of 30 from the theoretical limit.

12.3   SHIELDING

Capacitive sensors often work with sensor capacitance of 1 pF or less. Shielding the sensor plates from extraneous fields is critical for good performance. Even though a synchronous demodulator may reject interfering signals, the sensitivity of the input circuit may be so high that out-of-band signals may saturate the input amplifier. Shielding in general must attenuate both electric and magnetic fields, but because of the very high impedance of capacitive sensor inputs, electric fields are considerably more important. A strong magnetic field which couples an interfering signal to a circuit loop can affect sensor circuits; but, for sensitive PC board circuits, conductors can be paired and "twisted" using vias to avoid loops.

Electric shield effectiveness (S) is measured in dB by

$$S = 20 \ \log \frac{E_1}{E_2} \qquad\qquad 12.14$$

where E_1 is the incident field and E_2 is the attenuated field inside the shield. A conductive shield with thickness d attenuates electric fields by absorption A and by reflection R. The distance that the field penetrates into the shield is measured by skin depth δ, and absorption of a shield of thickness d is calculated as 8.69 d/δ so it increases by 8.69 dB (1/e, or 37%) with every skin-depth increase in thickness. The total shield effectiveness is the sum of this absorption and the attenuation due to reflection; when a plane shield is close to the source of the radiation the shield effectiveness S is

$$S = A + R = 8.69 \frac{d}{\delta} + 322 - 10 \ \log \left[f^3 r^2 \frac{\mu_r}{\sigma_r} \right] \qquad \text{dB} \qquad 12.15$$

d = thickness, m

δ = skin depth, m

f = frequency, Hz

r = distance from shield to source, m

μ_r = permeability of shield relative to vacuum

σ_r = conductivity of shield relative to copper

Although the absorption term A is independent of the distance of the radiator, the reflection loss measured by the second and third terms is not. The r^2 term is due to the change in the impedance of a dipole radiator, from very high near the dipole to 377 Ω at $\lambda/2\pi$ m away, where λ is the wavelength. The skin depth δ is calculated as

$$\delta = \sqrt{\frac{1}{\pi f \mu \sigma}} \quad m \qquad\qquad 12.16$$

For copper, $\mu = \mu_0 = 4\pi \times 10^{-7}$ and $\sigma = \sigma_c = 5.75 \times 10^7$ mho/m; eq. 12.16 becomes

$$\delta = \frac{0.0664}{\sqrt{f}} \quad m \qquad\qquad 12.17$$

At 25 kHz, the skin depth of copper is 0.420 mm. For 0.420 mm thick (one skin depth) copper and with $r = 1$ m, eq. 12.15 evaluates as

$$S = 8.69 + 322 - 132 = 199 \quad dB \qquad\qquad 12.18$$

or about one part in 10^{10}. With increasing frequency, the absorption term increases as $\sqrt{f}$ while the reflection term decreases, but at 250 kHz S is still 188 dB.

As the source distance increases past $\lambda/2\pi$, eq. 12.15 is inaccurate, and the far-field equivalent is used

$$S = A + R = 8.69\frac{d}{\delta} + 168 - 10 \, \log\left[f\frac{\mu_r}{\sigma_r}\right] \quad dB \qquad\qquad 12.19$$

The reflection loss of a copper shield in the far field varies from 150 dB at DC to 88 dB at 100 MHz. For more details on shielding effectiveness, see Ott [1988, pp. 159–200].

A graph of eq. 12.19 with $d = 0.42$ mm and $r = 0.1$ m shows the behavior of the two components of the shield attenuation (Figure 12.7).

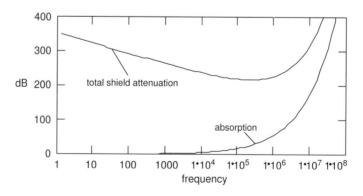

Figure 12.7   Shield attenuation vs. frequency

As shielding effectiveness is only fractionally lower with any reasonably conductive material, shield design is reduced to making sure that the sensor is completely surrounded with shield. If an opening is needed, forming a tube or channel instead of a simple cutout works better. A channel length-to-width ratio of five or more will attenuate low frequency incident fields about as well as the unopened shield.

12.4   GROUNDING

Noise reduction and low noise transistors can be sabotaged by a bad grounding scheme. A good ground, like security and freedom, cannot be won or lost; it must be won and rewon many times. A proper treatment of grounding is found in Ott [1988] and Morrison [1977], but a brief review is presented here for designs where analog capacitive sensors share a circuit board with digital logic.

12.4.1   Ideal ground

A poor ground is contaminated by voltage gradients caused by return currents from the analog circuit or from nearby digital circuits; a good ground is all at the same potential. Signal coupling from the same circuit can cause high frequency oscillation, low frequency oscillation (motorboating), or hysteresis, while coupling from a digital circuit can look like repetitive impulse noise with simple digital circuits or Gaussian noise for more complex digital circuits.

At first glance digital noise due to crosstalk may be indistinguishable from Gaussian preamplifier noise, but digital noise can be identified as it increases with frequency at about 6 dB/octave. This high frequency peaking is because digital noise generally couples through the reactive ground impedance or a reactive capacitive coupling; each of these coupling mechanisms results in 6 dB/octave spectral shaping. Digital noise is the most troublesome for low level circuits. Capacitive coupling is easily handled by separation, guarding, or shielding, but reducing ground coupling is more difficult.

There are three ways to establish an ideal unipotential ground: zero impedance, zero current flow, or correct connection of grounds.

Zero impedance

In any real circuit, some current must flow in the ground connections, so reducing the impedance of the ground connections as much as possible will improve performance. For capacitive sensors, which are much more sensitive to electric fields than magnetic sensors, a continuous ground plane under the preamplifier circuits is best. Failing that, a well-connected mesh of wires or circuit traces should be used. For low frequency signals, below, say, 20 kHz, the resistance of the plane is important, but at higher frequencies, skin effect predominates and any thickness over 0.03 mm is adequate. If a ground plane is used without special routing consideration for return currents, a 20 dB advantage over a two-layer board will be realized, but improvements past that figure will usually require rerouting of return currents away from the affected ground.

Zero current flow

Zero current flow can be approximated by ensuring that all signal paths are terminated in infinite impedance receivers. This will guarantee that no current circulates in the ground, and the ground will be ideal. This can be approximated by using high impedance receivers for all signals, but this technique fails at high frequencies where transmission lines need to be terminated properly for good signal integrity. Also, some loads may be low impedance, and providing a high impedance receiver for each load would add circuit complexity. Another factor which mitigates against zero current flow is the ground current caused by signals coupling to ground through parasitic capacitances.

Correct connection

Even when zero current flow cannot be arranged, correct circuit design can guarantee that currents do not flow in the low level analog ground system, or if they do, they flow through zero impedance connections. This is demonstrated by first analyzing a poor connection (Figure 12.8).

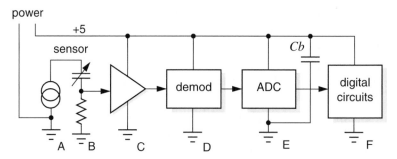

Figure 12.8   Poor ground connection

This circuit violates all the rules:

- No instructions for PC layout are included; the worst case PC ground connection, a single trace connecting grounds *A* through *F*, may result
- Power supply current to all circuits returns through connection *ABC*. A voltage drop equal to the ground resistance between *A* and *C* times the power supply current is added to the sensor voltage.
- Digital power current through bypass capacitor *Cb* flows through the ADC ground
- Analog circuits use noisy digital power

A better connection is shown in Figure 12.9. This circuit is considerably improved. No current flows through connection *A-B-C* to degrade the low-level analog performance except power supply returns from those circuits. The noisy digital +5 voltage has been filtered by *Rb-Cb2*, and *Cb* now keeps the digital power supply bypass currents in a small local loop. The ground connection is ambiguous, however, and should be formalized so the PC board layout is deterministic, as shown in Figure 12.10.

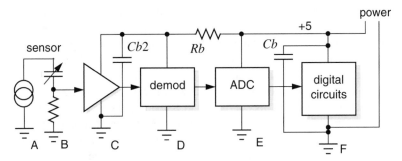

Figure 12.9 Better ground connection

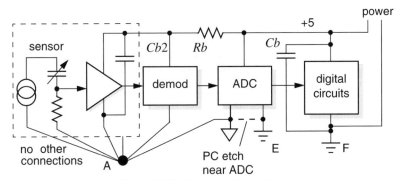

Figure 12.10 Best ground connection

Here the ground connections are unambiguous, an analog ground symbol is added so the PC layout process is forced to keep analog and digital grounds separate, a single ground node is used for the analog circuits, and a single connection between analog and digital grounds is specified. A grounded shield connects to the analog ground node.

12.5   STABILITY

Capacitive sensors are constructed with two or more conducting electrodes and an insulating support. These components will change size in response to environmental factors, and long-term stability may be a concern. These effects were briefly discussed in "Limits to precision" in Section 3.4, and the effect of environmental variables on air is found in "Dielectric constant of air" in Section 6.3.2. The stability of carefully constructed micrometers and tiltmeters [Jones, 1973] was seen in Section 9.13 to be about one part in 10^9 per day.

12.5.1 Temperature

Conductors

Most common metals and alloys have a temperature coefficient of linear expansion in the range of 9–32 μm/m/°C (5–18 μin/in/°F). Silicon, at 5 μm/m/°C, is quite stable, and specially formulated alloys such as Invar have even lower coefficients. Achieving the theoretical limiting displacement calculated above as 3×10^{-10} would require a stability of 3×10^{-4} µm/m. With a stable material such as silicon, a temperature change as low as 6×10^{-5}°C will cause an expansion equal to the limiting displacement if the plate spacing is determined directly by the silicon.

Invar [Harper, 1993, p. 5.44] is an alloy composed of 36% Ni, 0.4% Mn, 0.1% C, with the remainder Fe. Its coefficient of linear expansion between 0 and 38°F is 0.877 µm/m/°C, but at higher and lower temperatures its stability degrades to a pedestrian 10–15 µm/m/°C. Its coefficient is affected by annealing and cold working and may be artificially reduced to zero by these methods, but time and temperature will cause the coefficient to return to about 0.877.

Insulators

Insulators have a much wider range of thermal expansion than metals (Table 12.3).

Table 12.3 Temperature coefficient of some insulators

Pyrex glass	3.2 ppm/°C
Diamond	1.1
Fused quartz, SiO_2	0.5
Silicon nitride	2.0
Soda lime glass	9.2
Borosilicate glass	3.3
Acrylic	50–90
Epoxy	45–65
Polyamide	55
Polyimide	20
Polyester	30
TFE	100

Data from Harper [1970, p. 2.65] and *CRC Handbook for Applied Engineering Science* [1973, pp. 152–153].

For applications in high-temperature environments, TFE and polyimide have upper temperature limits of 260°C.

12.5.2 Stability with time

Long-term stability of materials is a concern for very sensitive applications. Standard meter bars machined from platinum-iridium alloy and bronze were found to change by three parts and 20 parts in 10^{-7}, respectively, in a 52 yr period. Gage blocks stable to one part in 10^{-7}/yr have been built from heat treated stainless steel and titanium carbide, and 70-30 annealed brass is also very good. Light alloys are unstable.

12.5.3 Stability with humidity

Metals have insignificant water absorption and chemical absorption and are generally quite stable to any chemical challenge other than acid. Most insulators, however, absorb water from the air during high humidity and expand. In the case of wood crosswise to the grain, this expansion is at its maximum, reaching 1/8 in per 1.5 ft in my kitchen cabinets. Glass and fused silica do not absorb water; a few plastics also do not absorb water, such as TFE (Teflon™) and polyimide (Kapton™), and some absorb very little, like mica. For the plastics, polyimide would be a good choice of insulator for critical capacitive sensor applications, with low temperature coefficient and low water absorption, excellent dielectric strength (5400 V/mil in thin layers), low dielectric constant (3.5), high volume resistivity (10^{18}), reasonably low loss tangent (0.003), and excellent tensile strength and abrasion resistance.

Another possible stability problem is in applications where insulators are exposed to chemicals or chemical vapor; for example, plastics like ABS or styrene will expand in the presence of organic solvents. Good choices for fluid chemical resistance are polyethylene, TFE, fluorinated ethylene propylene (FEP), and polyimide.

13

Hazards

Several hazards, unsuspected by practitioners of mainstream electronic engineering, await the unwary designer of capacitive sensors.

13.1   LEAKAGE

With small capacitive plate areas, electrode capacity is often a fraction of a pF. The silicon-based accelerometer of Chapter 15, for example, has a sensor capacitance of 0.1 pF. Low power circuit design requires a low excitation frequency, so the measured impedance of the capacitive sensor can often be many megohms, and measuring this impedance with good precision may require amplifier input impedance and printed circuit board and component leakage impedance to be many hundreds of megohms.

Leakage paths on the printed circuit board are usually a surface effect, not a bulk effect, and can be caused by:

- Rosin residue or lubricant acting as an attractant for dirt and dust
- Incompatible PC board rosin and cleaning fluids which chemically react to produce a conductive residue
- Conductive paint. Some black pigments are heavily carbon-particle loaded, and the paint will dry to a high-resistance conductor
- High humidity

13.1.1   Guarding against leakage

Surface leakage can be combatted by guarding. A typical high impedance guarded amplifier is shown in Figure 13.1.

The op amp input bias current may be as low as 10 fA. If a board leakage resistance path of 10^{15} Ω connects the input to a +10 V rail in the circuit above, the leakage current I_L will equal the op amp bias current and it will cause excess noise as well as shifts in the

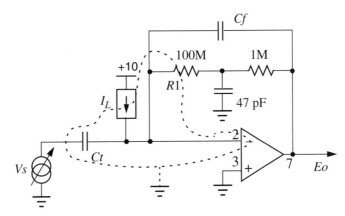

Figure 13.1   Guarding leakage paths, low-Z amplifier

bias point. It is very difficult to achieve this level of resistance on an unguarded PC board surface; very clean Teflon surfaces and long paths would be needed.

Surface leakage guarding

Luckily, surface guarding is a very effective technique to interrupt leakage paths. If a grounded printed circuit trace is added which surrounds the high impedance node, shown above as a dotted line, the leakage currents are interrupted. For surface mount construction, the guard trace must be on the same side of the PC board as the components; through-hole construction requires guards on both top and bottom surfaces. To guard pin 2 of the amplifier from leakage to pins 1 and 3, the trace should pass between pins 2 and 1 and also 2 and 3. For a low-Z or feedback amplifier as shown in Figure 13.1, the guard is grounded; for a high impedance amplifier the guard is connected to the amplifier output. Solder mask which covers the guard trace would produce a path which conducts surface currents over the guard, so solder mask should be relieved over the guard foil.

If $R1$ is a surface mount part, it can trap conductive residues under its body, or it can collect a conductive residue on its insulating body. One repair is to provide a solder-mask-relieved spot of grounded copper under the body of the part, but this method is unpopular with production personnel and also provides only a 90% solution, as surface currents can flow on the top surface of the part. For very high impedances or problem environments, somewhat better performance is available: $R1$ can be split into two series 50M resistors, with a guard surrounding the node where they connect (Figure 13.2). Capacitor leakage current is usually not a serious problem, as the DC potential across the components is generally near zero. For example, in the circuit shown in Figure 13.2, all the circuit's capacitors will have zero DC potential if the excitation Vs is centered on ground.

Bulk leakage guarding

Bulk PC board leakage currents in fiber reinforced boards can flow along impurities on the surface of the fibers. This is more of a problem for through-hole construction than surface mount. These currents can be interrupted by stitching small-diameter plated-through holes through the board in a way that will intersect all fibers, and connecting them to ground or guard voltage (Figure 13.3).

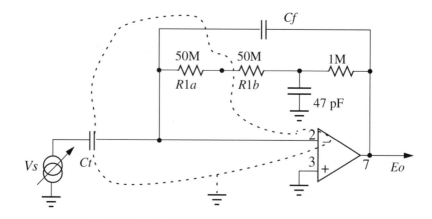

Figure 13.2   Guarding leakage paths, low-Z amplifier

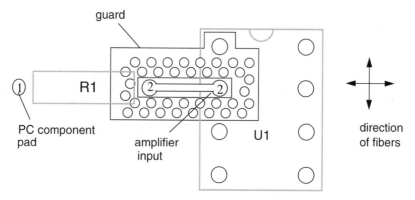

Figure 13.3   Guarding bulk PC leakage

Here, the sensitive node between a through-hole operational amplifier input, $U1$-2, and a high-value resistor, $R1$-2, has been guarded with etch, shown shaded, on both top and bottom sides. Many small via holes are stitched through the guard to interrupt bulk leakage current along the glass fibers. As the fibers are at 90° angles, the pattern shown is sufficient to interrupt all fiber-surface conduction.

Printed circuit board test patterns

Test coupons for measuring surface leakage can be used. These are interdigitated combs, specified by the Institute of Printed Circuits (IPC). The recommended minimum resistance is 10 MΩ. With normal PC board production techniques, the resistance should be many hundreds of megohms.

Some water-based fluxing processes may leave a white residue which has hundreds of ohms of surface resistance if incompletely cleaned. Keeping the board clean is also important, as surface impurities can be conductive, or can become conductive with high relative humidity. Nonconductive conformal coating can be used in extreme cases.

13.2   STATIC DISCHARGE

In position-measuring applications, capacitive sensors often have one moving plate. If the plate is coated with a dielectric such as solder mask, it can accumulate static charge by triboelectric charging. This is the familiar rub-the-comb-on-the-silk-shirt effect. When the accumulated charges reach a threshold voltage, the Paschen voltage (679 V in air at 1 atm), they can discharge in a small spark which can cause a large if mercifully brief transient in the nearby high impedance amplifier.

Paschen's law

Air breakdown at normal pressure happens when a free electron, propelled by a strong electric field within a narrow gap, dislodges enough other electrons in collisions with air molecules to produce a chain reaction. The final breakdown is a stream of positive ions which proceeds from anode to cathode [Javitz, 1972, p. 337].

Paschen's law can be shown, at large spacing, in Figure 13.4. A magnified view for low spacing or low pressure is shown in Figure 13.5.

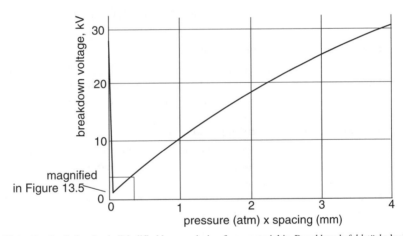

Figure 13.4   Paschen's law in air [Modified by permission from material in *Durchbruchsfeldstärke* by W.O. Schumann, pp. 51, 114, 1923. Berlin, Germany: Springer-Verlag GmbH & Co.]

At small spacing and low pressure, the mean free path of an electron at 10^{-4} mm is not long enough to produce many electrons by collision, and the breakdown voltage increases even though the field strength goes up. The population of free electrons to initiate the chain reaction is dependent on background ionizing radiation such as light and cosmic radiation and is invariant with field strength; no electrons are emitted from a conductor except at very high field strengths.

13.2.1 To combat static discharge

Raise the humidity

The surface leakage resistance of almost all insulators is is a strong function of humidity. At reasonable values of relative humidity, say 20% and higher, triboelectric discharge is not hazardous.

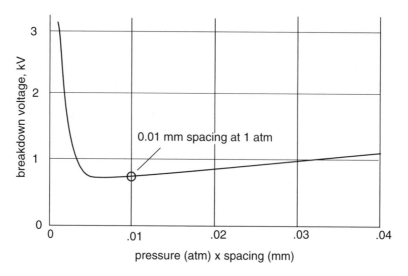

Figure 13.5   Paschen's law, detail

Use a static dissipation treatment

Many of these treatments are deliquescent and should be tested since a harmful conductive water deposit may form. Also, since most work by absorbing water vapor, the conductivity can vary by factors of 1000:1 when relative humidity changes from 5% to 95%.

Redesign the circuit to be zap-resistant

As an example of a zap-sensitive circuit, a plate counting circuit which keeps track of multiplate coarse position can pick up an extraneous count pulse during a static discharge, causing a large unrecoverable position error. The counter can be redesigned as shown in Figure 18.6 so that a single measurement which is different from its neighbors is assumed to be a discharge and is discarded.

Use bare metal plates without a dielectric coating

Bare metal will not have triboelectric-effect static discharge problems. Corrosion of bare copper is minimized by use of a corrosion-resistant metal plating such as nickel or gold over nickel.

Use similar materials

The triboelectric series listed in any physics reference shows which materials produce the strongest triboelectric charge when rubbed. The best case is to use identical materials, which have no triboelectric effect.

Ionize the air

Alpha particles generated by low-level radioactive materials produce charged ions by collision with air molecules, and static charge is dissipated by the slightly conductive air. Another technique is to excite a sharp point with high AC voltage, causing a corona discharge at the tip and launching a cloud of positive and negative ions into the air. One company that specializes in air ionization equipment is NRD, Inc., P.O. Box 310, Grand Island, NY 14072-0310.

13.3   STATIC CHARGE

Low level static charge can build up from the effects above. At levels where it is too low in voltage to spark, it may still cause problems. A static DC voltage on capacitive plates which are measuring an AC signal can turn into an AC voltage by $V = Q/C$. As Q is constant, a mechanical vibration which changes spacing and hence capacitance will produce an AC voltage V. If the system has mechanical resonances, this voltage will respond to a mechanical excitation impulse with a damped sinusoid voltage transient which can be many times larger than the signal being measured. Some fixes are listed below.

Use the methods listed above

All of the techniques for static discharge will also fix static charge problems.

Use a higher frequency carrier

Mechanical vibrations and resonances are usually in a low frequency range, say below 10 kHz. Use a 100 kHz carrier and a demodulator which includes a highpass filter to eliminate mechanical resonance and static charge effects.

Use adequate headroom in amplifiers

Even if the carrier frequency is well above the frequency of mechanical vibration, if the vibration is strong enough, the static-charge-induced voltage may exceed the linear range of the amplifier. Make sure adequate linear range is provided, or use a passive highpass filter which precedes the active amplifier.

13.4   CROSSTALK IN SILICON CIRCUITS

Silicon implementations of capacitive sensor circuits have different signal integrity problems than discrete implementations. All components on a silicon chip are capacitively coupled to the common substrate through either a silicon dioxide layer or a reverse-biased p-n junction. In addition, the p-n junction's capacitance is nonlinear, and temperature-dependent diode leakage current is added. The substrate voltage will not be the same as the reference voltage of the PC board the chip is soldered to, due to ground return currents, so care must be taken to bring the chip's signal ground reference out to sensitive input circuits through a dedicated pin which is not shared by load current or power return currents.

Other unwanted crosstalk is caused by parasitic capacitances between close-spaced signal and clock lines. Capacitively coupled clock frequency signals near the excitation frequency of a sensor will alias down to a harmful DC or low frequency signal at the demodulator output. The very high impedance nodes and connection wires of capacitive sensors can be shielded from adjacent clock interconnects with an interposed dummy grounded line or a guard ring as in PC board practice, and a low-impedance polysilicon plate or diffusion well can be layered underneath [Hosticka, 1990, p. 1348].

The use of differential paths and differential circuits for signal conduction and generation offers an improvement over traditional single-ended design. With careful symmetric layout, differential paths reject crosstalk, improve power supply coupling, and increase dynamic range by doubling signal level while keeping noise constant. Moving the differential partition back to the sensors, if possible, can first-order compensate for several types of sensor inaccuracy and environmental dependence.

14

Electret microphone

Electret microphones are the only application covered in this book which do not use high frequency excitation. A DC bias is used so the circuits are considerably simplified. The price paid for this simplicity is the poorer noise performance, as two mechanisms contribute: the increased low frequency noise of semiconductor junctions and the much higher electrode impedances, as a capacitor's impedance is a direct function of frequency. These drawbacks have been solved by manufacturers of electret microphones by integrating a high input impedance JFET amplifier with the microphone and using a high voltage DC bias on the order of 100–500 V for more sensitivity.

Electret microphones cost less than $1 and have good audio performance. They are available from a variety of sources such as Panasonic and Primo.

14.1   ELECTRET MICROPHONE CONSTRUCTION

An electret microphone is shown in Figure 14.1. The diaphragm may be metallized plastic 20 μm thick. Vents in the electret allow free diaphragm movement, and a small vent in the back of the case allows enough air movement to equalize pressure, but not enough to spoil the low frequency response. The electret is a plastic material, originally Carnauba wax, now often polytetraflouroethylene. The electret traps a charge when biased with a high DC voltage at high temperature and retains it when allowed to cool, with a charge retention time of 2–10 yr. The temperature coefficient of sensitivity is not as good as other microphone types due to electret dielectric constant change. The surface voltage can reach 2 kV, but is kept below Paschen's threshold to avoid air breakdown (see Section 13.2).

Typical electret microphone specifications

Directionality .... > 10 dB
Design load impedance .... 2 kΩ +/– 20%
S/N ratio............................ > 40 dB at 1 kHz, 1 mbar, A weighted

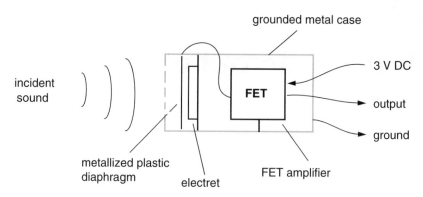

Figure 14.1   Electret microphone

Operating voltage .... 1.1–9 V
Circuit current .... 250 mA max
Signal .... –67 dBV signal at 1 µbar
Noise .... –110 dBV noise in audio range

Circuit description

The FET is typically an n-channel J-FET, such as type 2SK997 made by NEC (Figure 14.2). The microphone is specified with a 2 kΩ load resistor so the circuit will operate down to the 1.5 V provided by battery operation, but the gain increases with load resistance with no performance penalty and noise will decrease slightly. These microphones have a maximum quiescent current on the order of 250 µA, so the maximum resistor which can be used in a 10 V circuit is 10/250 µA or 40 kΩ.

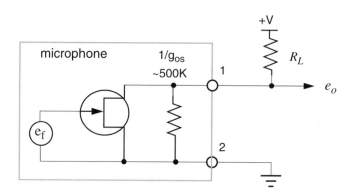

Figure 14.2   Equivalent circuit

The voltage e_f is the sum of a signal source, e_s, and an equivalent input noise component, e_n. The microphone diaphragm is approximately 0.5 cm in diameter. The signal voltage e_s is produced on the capacitance formed by the moving diaphragm and a fixed plate, by the relationships $V = Q/C$ and $C = \varepsilon_0 \varepsilon_r A/s$, where Q is the charge produced by the electret and C is the capacitance, inversely proportional to spacing s so that

$$e_s = \frac{s}{s_{avg}} \cdot V \qquad\qquad 14.1$$

with s_{avg} the average spacing. The FET input capacitance is 5 pF, smaller than the 10 pF diaphragm capacitance.

The output noise of the 2SK997 JFET is 6 μV into a 2 kΩ load impedance with A weighting. This is about 42 nV/$\sqrt{\text{Hz}}$, compared to op amp noise of 4–40 nV/$\sqrt{\text{Hz}}$, so the FET noise will normally dominate the output SNR and a low noise type amplifier need not be used. The 2SK997 FET also lists these specifications:

Voltage gain into 2K .... –4.5 dB/+0.5 dB
V_{DS} .... 20 V
I_{DSS} .... 120–300 μA
g_{OS} .... 850 μmho typ

14.2   EQUIVALENT CIRCUIT

A block diagram of the equivalent circuit is given in Figure 14.3. A 10 kΩ resistor would have 1.27 μV rms noise at a 10 kHz bandwidth, and 12.7 nV/$\sqrt{\text{Hz}}$ noise. Actual resistors have an excess noise which is voltage-dependent and changes with construction, but the excess noise of carbon film resistors should be insignificant in this circuit. For more on resistor noise, see Section 10.4.

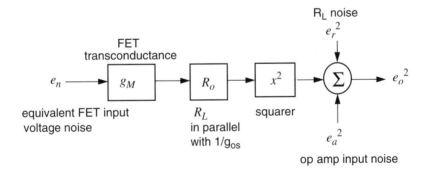

Figure 14.3   Block diagram

$$e_o^2 = \left(e_n g_M \frac{R_L}{R_L g_{os} + 1} \right)^2 + e_r^2 + e_a^2 \qquad\qquad 14.2$$

Resistor noise $e_r^2 = 4kTRB$

$4kT = 1.64 \times 10^{-20}$ at 25°C

B = bandwidth, in Hz

Note that the output signal voltage e_o increases linearly with R_L for $1/g_{os} \gg R_L$, but the output noise voltage contribution of R_L increases as the square root of R_L, so each time R_L is quadrupled the noise contribution of R_L to the SNR is halved. SPICE models show a limit to this improvement due to two factors:

1. Running out of supply voltage, at $R_L > V_{supply}/250\ \mu A$, or 40 k$\Omega$ (10 V). If an active current source is used, this limit is extended
2. The shunting effect of the FET output conductance, about 25 kΩ

The microphone terminal characteristics at DC will be those of a JFET with zero bias. If $V_{cc} = 6$ V, the load line can be drawn (Figure 14.4).

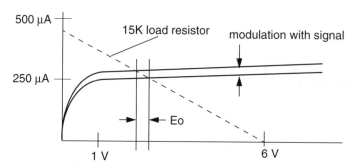

Figure 14.4   Electret microphone load line

At low voltages, the terminal characteristic is a variable resistance, and at high voltages, the characteristic is that of a modulated current source. At 10 kΩ load resistance, the supply voltage must be more than 2.5 V (R_L) + 1.5 V (FET) or 4 V so the load point stays in the linear modulated-current-source region of the FET.

14.2.1 Miller capacitance

Miller capacitance due to FET drain to gate capacitance Cdg plus stray capacitance contributes a response pole (Figure 14.5).

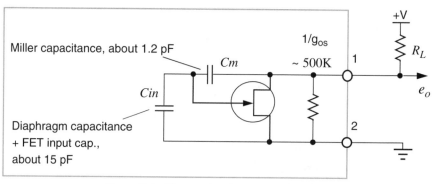

Figure 14.5   Electret microphone equivalent circuit

Since the gate circuit has a very long (seconds) time constant, limited only by small gate leakage currents, the effect of Miller capacitance is broadband instead of its normal high-frequency-only shape. It will reduce the *Rds* impedance as measured at the output by feeding a portion of the drain voltage change back to the FET gate, and the resultant negative feedback decreases *Rds* proportionately to C_m/C_{in}. This results in a reduction of *Rds* from 500 kΩ at DC to about 20 kΩ at audio frequencies. This lower value of *Rds* will shunt the load resistance and decrease the maximum available voltage gain from the FET. The *V-i* curve at the electret microphone output, measured for two devices at a frequency well below the Miller-capacitance induced response pole, is shown in Figure 14.6.

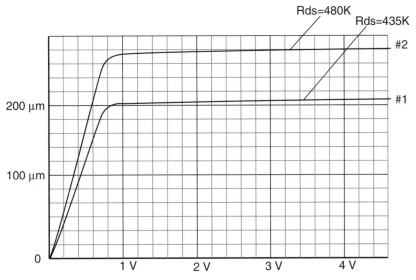

Figure 14.6   Measured electret microphone curves

Note the large increase of output resistance compared to the 25 kΩ high frequency value.

15

Accelerometer

A good example of the use of capacitive sensors for silicon implementation is Analog Devices' surface-machined accelerometer.

Integrated circuit designers use precision lithography and micromachining to produce micron-dimension sensors in silicon. Several different silicon accelerometers have been built with bulk micromachining and piezoresistive sensors beginning in the 1970s, but the large size and the large number of process steps of the bulk technique as well as the temperature sensitivity of the piezoresistive sensors have slowed commercial acceptance. More recently, capacitive position sensing and electrostatic force-balance feedback have been used in a surface-machined device which uses more conventional integrated circuit processes. Surface micromachining allows a more highly integrated design with much smaller chip dimensions.

Analog Devices' ADXL50 is the first commercially available surface-microma-chined accelerometer with integrated signal processing. It measures acceleration in a bandwidth from DC to 1 kHz with 0.2% linearity, and it outputs a scaled DC voltage. It is fabricated on a 9 mm^2 chip. It also is a force-balance device which uses electrostatic force to null the acceleration force on the "proof" (seismic) mass, with advantages in bandwidth, self test, and linearity.

15.1   ACCELEROMETER DESIGN

15.1.1   Surface micromachining vs. bulk micromachining

[Riedel, 1993, pp. 3–7, used with permission]

Micromachining is a processing technique used to manufacture tiny mechanical structures from silicon. A silicon wafer of the type used to make semiconductors can be etched to produce small beams, masses, gears, and other structures measuring only a few thousandths of an inch.

Micromachining comes in two varieties: surface and bulk. Prior to the ADXL50, all the available micromachined devices used bulk micromachining. It was discovered in the 1950s that acid solutions attack different planes of crystalline silicon at different rates, depending on the crystal orientation. By exposing an area of silicon with a specific crystalline structure to acid, cavities with precisely angled walls are created.

Bulk micromachined accelerometers have existed for several years. Typically they consist of a membrane or diaphragm of silicon, roughly 10 μm thick, that is vertically formed in the wafer by chemical etching. In the center of the membrane is a large mass of silicon. On the top surface of the device, near the edge of the membrane, thin-film piezoresistors sensitive to strain and deformation are deposited. Most of the membrane is removed, leaving tethers with these resistors suspending the central mass. Vertical acceleration causes the test mass to move, deforming the diaphragm and changing the resistance of the piezoresistors. Bulk micromachined devices are large by IC standards—about 20× the size of the surface-machined ADXL50. Large size, coupled with the fact that the process for manufacturing bulk micromachines is inconsistent with semiconductor-circuit fabrication techniques, requires that signal conditioning be off-chip. Bulk piezoresistive accelerometers are very sensitive to temperature effects and difficult to test fully.

Surface micromachining, a more sophisticated technique than bulk micromachining, creates much smaller, more intricate, and precisely patterned structures. It adapts manufacturing techniques perfected for making ICs to produce mechanical structures close to the surface of the silicon substrate. Chemical machining is accomplished by deposition, then etching multiple thin films and layers of silicon and silicon-oxide to form complex mechanical structures. The feature dimensions of surface micromachined devices are typically 1 to 2 μm, similar to the feature dimensions of conventional electronic circuits. Most importantly, surface micromachining lends itself to the inclusion of conventional electronic circuitry on the same die. Thus, the surface-micromachined ADXL50 includes signal conditioning, resulting in a fully scaled, referenced and temperature-compensated volt-level output. Surface micromachining leverages the cost economies of standard IC wafer processing techniques, producing a highly integrated product at low cost.

15.1.2 Force balance

Early accelerometers measured the displacement of the proof mass. Problems with this approach are the presence of a resonant peak, the difficulty of achieving good dynamic range, and the need to carefully calibrate the force vs. displacement relationship. Force balancing allows the proof mass to deflect only microscopically, detects and amplifies this displacement, and feeds back a force to restore the rest position. With a high gain amplifier, the mass is nearly stationary and the linearity is determined by the linearity and precision of the voltage-to-force transducer rather than the suspension characteristics.

15.1.3 Capacitive feedback in silicon accelerometers

Although piezoresistive sensing of the displacement of the proof mass was used for early devices, a recent survey [Yun, 1991, p. 204A-4] showed that the excessive temperature sensitivity of piezoresistivity was incompatible with the preferred force balance technique. Capacitive sensing has replaced piezoresistive sensing in six new designs cited by Yun [1991]. Sze [1994, pp. 192–193] lists these advantages of capacitive sensors over piezo-electric or piezoresistive types:

• Wide temperature range
• Low temperature coefficient
• High sensitivity
• Response to DC
• No zero shift due to shock

Bulk-machined capacitively sensed accelerometers can have excellent specifica-tions, as shown by a device built by CSEM with 1 μg resolution and cross-axis sensitivity of 0.4%, and Triton/IC Sensors' 0.1 μg, 120 dB dynamic range unit with cross-axis sensi-tivity of 0.001%. But fabricating the Triton design requires 27 separate lithography steps on five wafer surfaces of three bonded wafers.

15.2   ADXL50

Analog Devices' ADXL50 uses an exceptionally small (1 mm^2) capacitive sensor ele-ment. The full-scale range is ±50 g, compatible with automotive airbag deployment requirements, and the accuracy is 5% over temperature and power supply extremes. It uses a 5 V supply, and includes a calibrated high level output and a self testing feature, at a high volume price approaching $5.00. A more sensitive version, the ADXL05, spans the range of ±5 g with 0.005 g resolution and a typical nonlinearity of 0.3% full scale.

Two different photolithographic exposure processes are used to maximize process throughput: the capacitor fingers are exposed with fine line optics and the standard 4 μm BiCMOS signal processing circuits are exposed with the more relaxed optics of 1× lithog-raphy. The fine line exposure for the capacitive plates allows a very small gap width, which results in higher signal strength as the signal is inversely proportional to gap. The parallel connection of 42 similar 2 μm thick electrodes further increases the signal. The silicon-micromachined electrode design is a differential 0–180° type with square wave drive voltages (Figure 15.1).

The H-shaped proof mass and sensor electrode measures 500 μm × 625 μm. It is supported on the corners by slender tethers, measuring 2 μm square and 200 μm long, which anchor the central elements to the substrate and provide the electrical connection. The tethers are formed from crystalline silicon, a very stable and reliable spring element, and the geometry allows the central element to move only in the x-axis when displaced by acceleration. The strength of the tethers is sufficient to allow the device to withstand 2000 g of physical shock. Projecting from the central bar are 42 moving plates, of which four are illustrated. Each moving plate is adjacent to one fixed plate driven by 0° and one driven by 180° as shown, so that the capacitance to one plate will increase with displacement as $Cx_0 / (x_0 + \Delta x)$ and the capacitance to the other plate will decrease as $Cx_0/(x_0 - \Delta x)$, with x_0 the undisturbed position and Δx the displacement.

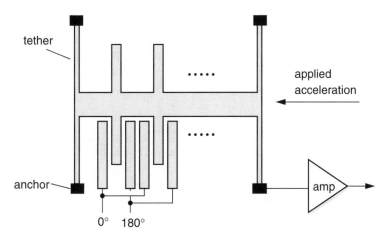

Figure 15.1   ADXL50 sensor electrodes

Sensor characteristics

The ADXL50 block diagram is shown in Figure 15.2. The sensor has approximately these characteristics:

Gap x_0.....................................2 µm

Electrode width W.................2 µm

Electrode total length L......... 100 µm × 50 × 2 = 10 mm total

Moving plate mass.................0.1×10^{-6} g

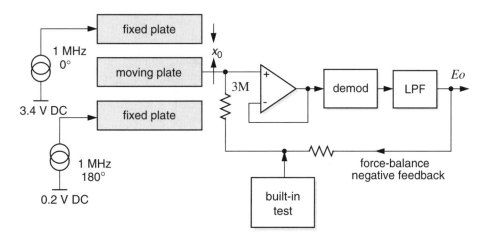

Figure 15.2   ADXL50 block diagram

The total sensor electrode capacitance, neglecting fringe fields, is $8.854 \cdot 10^{-12} \times A/d$ or about 0.1 pF. This small value would require extremely low amplifier input capacitance for accurate open-loop sensing, but with closed-loop operation the error contribution due to variations in gain is negligible if the amplifier gain is high. The open-loop change in capacitance with a 50 g acceleration is 0.01 pF, and the system can resolve a change of 20 aF, $20 \cdot 10^{-18}$ F, corresponding to a beam displacement of $20 \cdot 10^{-6}$ μm.

The amplifier is conventional, and the demodulator is a standard synchronous demodulator implemented in bipolar technology. The lowpass corner frequency has a single-pole response shape controlled by an external capacitor. The output signal is 1.8 V DC for zero acceleration, and spans ±1 V for ±50 g input.

Electrostatic restoring force

The acceleration force of a 0.1×10^{-9} kg proof mass at 50 g is 49×10^{-9} N. With a bias level of 1.8 V, midway between the two fixed plate DC levels, the electrostatic force is balanced. As the proof mass is deflected by acceleration, Eo provides a restoring voltage. The electrostatic restoring force F is developed by biasing the fixed plates with different DC levels and changing the DC level on the moving electrode through the 3 MΩ load resistor. From the electrostatic force equation, eq. 2.3 in Chapter 2, with $Vr = 1.6$ the maximum available electrostatic force is

$$ F = \frac{4.427 \cdot 10^{-12} \cdot \varepsilon_r A V^2}{x_0^2} = 68 \cdot 10^{-9} \quad \text{N} \qquad 15.1 $$

and is more than enough to balance the acceleration force. The nonlinearity with gap size x_0 is unimportant, as the gap does not change appreciably in operation.

Self test

When a logic signal is applied to the self test input, a voltage pulse is injected through the 3 MΩ load resistor to deflect the proof mass. If the system is working correctly, a proportional output voltage is produced.

16

StudSensor

The StudSensor (Figure 16.1) is another application that demonstrates the utility of capacitive technology. This low cost device (under $30) is a simple and effective use of capacitive sensors in a consumer product. As developed by Zircon International, Inc., the StudSensor helps locate the position of a wall stud (typically 2 × 4 or 2 × 6" wood) behind wallboard or plaster walls. Stud location is important for carpenters or homeowners attaching shelves to a wall or installing electrical outlets. Previously, wall studs could sometimes be located by banging on the wall with a hammer, but this method is somewhat destructive and not too reliable; happily, wall studs have a dielectric constant which is different from air's, and their location is revealed to a capacitive sensor.

The StudSensor, invented by Robert Franklin and Frank Fuller, was awarded U.S. Patent 4,099,118 on July 4, 1978.

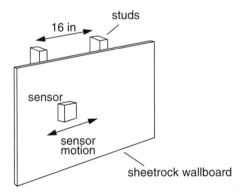

Figure 16.1   StudSensor in use

231

16.1   BLOCK DIAGRAM

Using the basics described earlier in this book, a relatively straightforward sensing circuit may be designed. The circuit is built with the familiar elements shown in Figure 16.2.

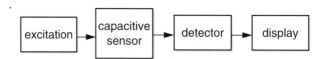

Figure 16.2   StudSensor block diagram

16.2   CIRCUIT DIAGRAM

In this capacitive sensor implementation (Figure 16.3) the electrode excitation is handled by a multivibrator, the integration circuit is a simple "leaky capacitor," and the measurement circuit is a voltage follower driving a display. The display shows the relative dielectric constant in the vicinity of the StudSensor's electrodes, which increases with the presence of a wall stud. The device is placed on a wall and the bias is adjusted so that the magnitude output just begins to register on the meter.

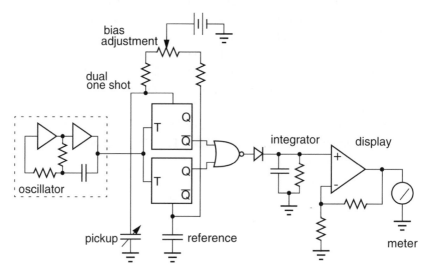

Figure 16.3   StudSensor circuit

Although this basic circuit accomplishes the task of displaying the change in dielectric constant, it can be improved for cost and usability. Three enhancements are included in the production StudSensor:

- Low cost LED display replaces meter
- Stud location sensitivity increased
- Self calibration circuit added

16.2.1 Low cost display

A numeric voltage display could be used for an absolute readout, but for this application only a relative response is needed. A simple magnitude display uses several voltage comparators and LEDs (Figure 16.4).

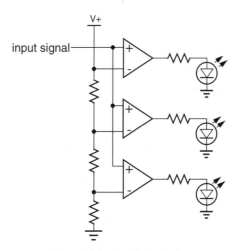

Figure 16.4   Magnitude display

16.2.2 Increased sensitivity

Increased sensitivity may be achieved through the use of two additional plate sensors. The single plate implementation has a relatively broad response and also is dependent on the stability of a reference capacitor. The addition of two additional capacitive sensor plates working in opposition to the primary plate increases the sharpness of the response curve. This connection also improves circuit stability, as two similar-dielectric capacitors are balanced.

The two additional plates are mounted beside the primary sensor (Figure 16.5), which then increases the x-axis sensitivity (Figure 16.6).

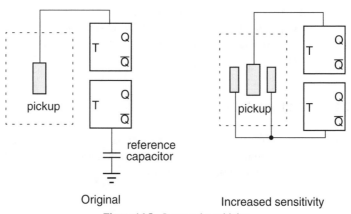

Figure 16.5   Increased sensitivity

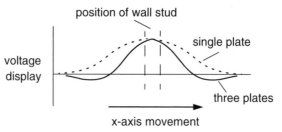

Figure 16.6   Sensitivity curve

16.2.3 Self calibration

One other design problem is the need to calibrate the sensor offset voltage to the wall-board thickness. The circuit above manually calibrates sensor offset with an adjustable bias control. This function can be automated with a FET-driven bias calibration circuit (Figure 16.7).

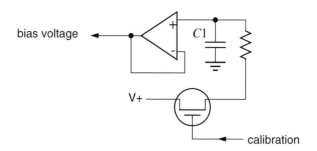

Figure 16.7   Bias calibration circuit

In the original circuit, the device must be placed on the wall away from a stud and manu-ally adjusted so the meter is in range. The new circuit initiates a calibration cycle on pow-erup which slowly charges $C1$ and increases the bias voltage until the first display LED indicator turns on. At that point, the calibration signal is turned off, the bias voltage stabi-lizes, and StudSensor is ready for operation.

In the event the device is coincidentally switched on when directly over a stud, the device will give no stud indication as it is moved. Recalibration over a new wall area cor-rects the problem.

16.3   MODIFIED CIRCUIT DIAGRAM

By making the adjustments described in section 16.2, it is relatively simple to add a low cost display, increased sensitivity, and self calibration to the original circuit. Figure 16.8 shows a diagram of the improved circuit.

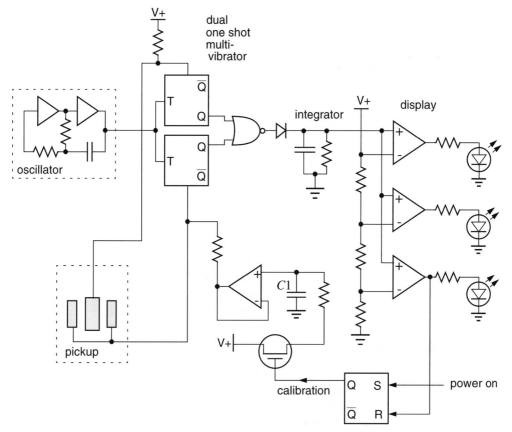

Figure 16.8   Modified circuit

17
Proximity detector

A single-plate capacitive proximity detector has been designed and breadboarded to develop the circuit for a low cost detector. The detector has good detection range and small size, and provides a threshold output as well as an analog output which is proportional to changes in capacitance to local objects.

17.1 BLOCK DIAGRAM

Figure 17.1 shows a block diagram of the circuit. The circuit uses two sense electrodes, A and B, with the capacitance Cs varying with target position. A bridge circuit is used so that component variation will cancel as much as possible; Cr is the reference plate capacitance. When $Cr = Cs$ the bridge is balanced and the amplifier output is near zero. Temperature and humidity variation effects are canceled if Cr's construction closely matches Cs's, so both are air-spaced construction. The 1.5×2.5 cm electrodes have a capacitance to ground of approximately 1 pF.

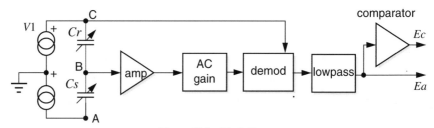

Figure 17.1   Block diagram

17.1.1 Lossy capacitance

The block diagram shows Cs and Cr as pure capacitors. More generally, Cs and Cr should be drawn as lossy capacitors (Figure 17.2).

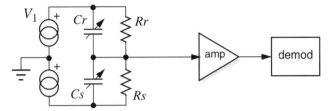

Figure 17.2 Lossy capacitors

The shunt resistance represents the dielectric loss tangent. Rr and Rs are insignificant for air-dielectric capacitors, but if the sensor environment includes lossy dielectrics like wood, they may become important. If Rr and Rs are not equal, and assuming sine wave excitation, the effect is to add a quadrature sine wave at 90° to the bridge output. This sine wave does not affect the demodulated output at low levels, since the demodulator circuit and lowpass combination rejects 90° components, but for maximum sensitivity a high gain AC amplifier is needed and the 90° component may saturate the amplifier. This is cured by adding a trim resistor Rq to the bridge to correct the $Rr - Rs$ difference, or by adding a second demodulator sensitive to 90° (and insensitive to 0°) along with a feedback circuit to automatically null the 90° output component.

17.1.2 Construction

With this construction (Figure 17.3), Cs, the external capacitance which will be modulated by target position, is balanced in a bridge circuit by Cr.

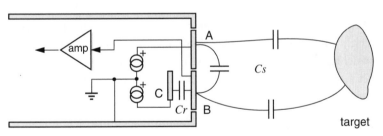

Figure 17.3 Cross section

Monopole response

The face view of the device is shown in Figure 17.4. With this electrode pattern, the response shape will be a nearly uniform sphere, with a small bias toward targets positioned in the direction of the face.

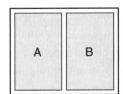

Figure 17.4 Face view, monopole response

Dipole response

If the shielded reference electrode C is instead brought up to the face, a dipole response results (Figure 17.5). Now, Cr and Cs will be equal for any target position which bisects the face in the vertical plane, producing a dipole response for horizontal motion (Figure 17.6).

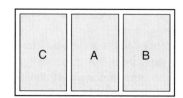

Figure 17.5   Face view, dipole response

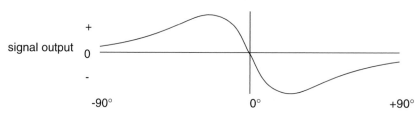

Figure 17.6   Dipole response

This presents the interesting option of adding two more electrodes in the vertical plane which can produce a similar dipole response for vertical motion. Then, with proper switching of electrodes and appropriate amplitude control of the excitation voltage, a target's approximate angular extent and the centroid of its vertical and horizontal position can be determined.

17.1.3 Carrier

The simplest choice of carrier is a square wave in the 10 kHz – 1 MHz range. As the carrier frequency is increased, lower reactance of the sensor capacitance helps minimize amplifier input current noise contribution. A second advantage of increased carrier frequency is that the injected building ground noise near the carrier frequency will decrease, as building ground noise is generally 60 Hz and harmonics and harmonic energy decrease with increasing frequency. Additional discussion on building ground noise can be found in Section 6.3.1.

Carrier frequency cannot be arbitrarily increased to avoid power frequency components, however, as higher frequency, high current amplifiers would be needed. Another factor to be considered in carrier frequency choice is the increasing use of fluorescent lamps which are powered by switching supplies in the 20 kHz – 200 kHz range. For example, the circuit shown in Figure 17.7 uses a 100 kHz carrier, well above power frequencies, but not difficult for conventional operational amplifiers to handle.

17.2   SCHEMATIC

Figure 17.7 shows a schematic of the proximity detector.

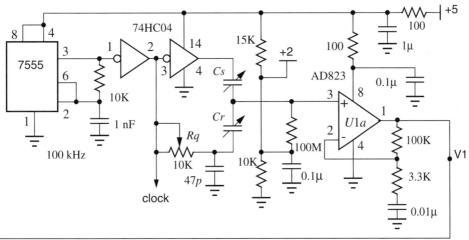

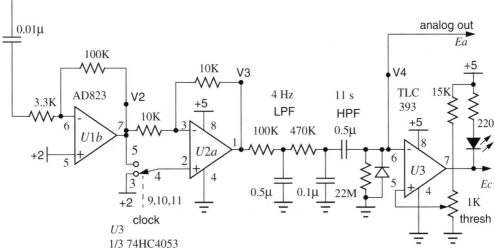

Figure 17.7   Proximity detector schematic

17.2.1   Excitation

A 7555 timer chip generates a square wave clock which is buffered by CMOS inverters to produce a 0° – 180° drive waveform which pulls accurately to the voltage rail. One output feeds Cs, the capacitance to be measured, and the other feeds balancing components Rq and Cr which are adjusted for null.

17.2.2   Input amplifier

For a sensor capacitance which is much less than the amplifier common-mode input capacitance, performance is improved by adding the *Vee* bootstrap circuit as in Figure 1.9

and increasing or bootstrapping the input resistor. For large, high capacity electrodes, the simpler circuit shown will be adequate with the input time constant contributing a useful low-cut filter. The amplifier input capacitance with careful construction will be on the order of 10 pF, which will cause about 20 dB gain loss with the 1 pF sensor capacity. A FET-input integrated amplifier with low current noise is shown, but 2–5× improvement in amplifier noise can be had by using a discrete FET preamp. The amplifier f_T, 15 MHz, limits high frequency response to about 0.5 MHz at 30× gain.

17.2.3 Demodulator

The full-wave demodulation shown in the schematic is preferable to half-wave demodulation as it rejects any remnant low frequency noise components which get by the input highpass circuits.

$U2a$ is connected as an inverter ($V3 = -V2$) with the switch as shown and as a follower ($V3 = V2$) with the switch in the opposite position. This produces a synchronous full-wave demodulator. If the phase shift through the signal path is close to zero, the voltage at $V3$ is a chopped waveform with an average DC level proportional to the input AC level with polarity controlled by input phase.

Lowpass, highpass filter

The crude 4 Hz lowpass filter removes the chopping components and also any high frequency in-band noise components, and the 11 s highpass filter rejects very slow DC variations which may be caused by component drift. High value resistors are used to keep filter capacitor size down, so the usual high impedance printed-circuit-board construction techniques should be used. The highpass filter may not be needed for less sensitive applications.

Comparator

A comparator compares the DC output to a preset threshold, and a LED diode shows when the threshold has been exceeded.

Null

For maximum sensitivity, the proximity detector is adjusted so that a null amplifier output is achieved with no object in the field. For a good null:

* Balance the fundamental output by adjusting Cr
* Balance the quadrature output by adjusting Rt
* Repeat as needed

If the amplifiers had infinite headroom, the goodness of null would not be too critical as the synchronous demodulator responds only to the in-phase fundamental component. With finite headroom, however, other null components such as quadrature and harmonics may be large enough to drive the final amplifier to clipping and compromise the demodulation.

With square wave excitation and wideband amplifiers, achieving an accurate null is more difficult, as all frequencies within the passband must be equalized in amplitude and phase. With either a sine wave drive or with a narrow bandpass filter, only one frequency need be considered. The circuit above uses a square wave drive, but each amplifier acts as a bandpass filter in the 5–500 kHz range with the upper frequency cutoff due to the amplifiers f_T. On the bench, the null at the amplifier output suffered from a quadrature component due to the small 20 ns delay between the 0 and 180° 5 V square wave drives, which was compensated by adjusting Rq.

Gage factor

The gain, or gage factor, is not too critical for a proximity detector. The circuit above is not accurate, as the phase shift through the amplifiers causes an uncompensated gain error in the demodulator output. A second similar amplifier circuit could be used to feed the carrier input to the demodulator to equalize delays, but this approach is not much better because of the extreme sensitivity of the phase shift to the filter center frequency and R and C values. Another approach would be to use a bandpass filter which has less phase shift, such as an FIR (finite impulse response) type, or a filter with more stable phase shift such as a crystal or ceramic resonator.

17.2.4 Power supply

The power supply can be any standard 5 V supply, but should be low noise, less than 10 mV. Note that most voltage regulators have high noise (over $100\ \mu V/\sqrt{Hz}$) and low output impedance measured in milliohms, so a simple capacitor does not do a good job at removing regulator output noise. A $0.005\ \Omega$ regulator output impedance with the normal 10 μF filter capacitor produces a 50 ns time constant, nowhere near long enough for useful filtering, and the current to slew a 10μ capacitor 5 mV at 50 kHz carrier frequency is only about 10 mA. Adding the $100\ \Omega$ resistor as shown reduces power supply noise to a few $\mu V/\sqrt{Hz}$.

17.2.5 Noise and stability

Although the amplifier used was not a particularly low noise device, the circuit's noise bandwidth is very low and noise was not a limiting factor. The amplifier output noise at $V3$ is equal to the input noise ($25\ nV/\sqrt{Hz}$ in a 500 kHz bandwidth) times the amplifier gain of 1000, or 18 mV rms. Power line noise at 60 Hz was much larger than the amplifier noise at the input stage, but it was effectively removed by the bandpass amplifiers.

The 18 mV noise component at the amplifier output is attenuated by the lowpass filter following the demodulator. The lowpass corner frequency is about 5 Hz, so the attenuation is the ratio of 500 kHz to 5 Hz or 100,000:1 for a filtered noise of 0.18 μV.

Stability was a much more serious problem. The demodulated DC output at the lowpass filter was affected by temperature, small movements of electrical components, and even air currents. At null, the amplifier stability is not a factor, so the presumed cause of the instability is the rather unstable electrode construction of glued-up printed circuit board stock and the unattached hookup wire electrode connections. The lowpass/highpass filter configuration rejects slow DC variations, however. If a true DC response were necessary, more stable mechanics and a better solution to the bandpass phase response problem would be needed.

Layout

The general layout shown in Figure 3.9 was used. Standard AM radio IF construction rules were followed, with the amplifiers in a linear array so the output could be as far as possible from the input, but with maximum amplifier gain, capacitive feedback from amplifier output to input caused oscillation at about 20 kHz. A copper ground plane fixed the problem.

17.3   PERFORMANCE

The limiting detection distance of the experimenter's hand was about 250 cm. As this was determined by stability rather than noise, and as noise was at least a factor of 1000 lower than limit signal, the range with careful construction could be extended by 60 dB. As the response of 2 cm electrodes (close to our 1.5×2.5 cm electrodes) is seen in Figure 2.39 to decrease by 20 dB for each 18 mm of separation, a 60 dB sensitivity improvement results in a detection range increase to only 304 cm.

An improved version would concentrate on improved mechanical stability. Also, the tradeoff between increasing predemodulation gain (better DC stability, but added risk of saturating the amplifier with quadrature signals or added noise) and increasing postdemodulation gain should be carefully evaluated for each application.

18

Vernier caliper

18.1   6 in CALIPER

A typical linear plate-counting system is illustrated by L.S. Starrett's model 723 vernier caliper (Figure 18.1).

This device uses sine-cosine pickup, coarse plate counting logic, and analog interpolation between plates to provide a resolution of 0.0001". The 723 is a general purpose digital-output LCD-display 6 in caliper, with a small electronics module $2 \times 0.6 \times 1.4"$, and it is specified at one year battery life, with two small silver-zinc cells used for power. A two-chip circuit is used, with a 5 μm CMOS ASIC for both drive and demodulation, using switched capacitor methods, and a 4 bit microprocessor handles math and display.

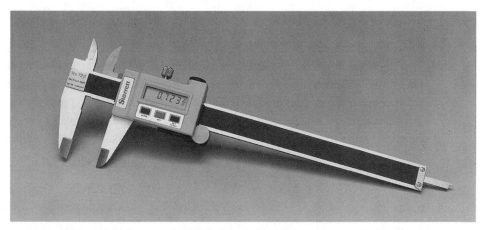

Figure 18.1   Starrett model 723 (U.S. patent 614,818, L. K. Baxter and R. J. Buehler, assigned to L. S. Starrett Co.)

243

The model 723 uses capacitive sensing to achieve the high resolution and good battery life. It uses multiple plates on 0.1 in centers, with a digital coarse plate counter keeping track of the position of the slider on the 60 plates and an analog sin-cos fine position sense for the 0.1% resolution. Its plate geometry is shown in Figure 18.2.

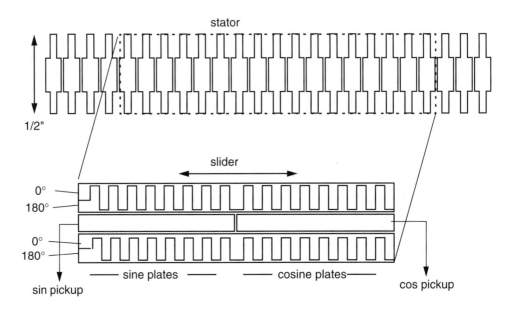

Figure 18.2   Starrett 723 plate geometry

A partial section of stator is shown. The 6 in long stator is fabricated from 1/32 in FR-4 printed circuit board with 1 oz copper and a laminated 0.006 in glass-epoxy cover layer. It is glued in a channel of the stainless steel caliper bar.

The slider plate is 1/16 in FR-4 printed circuit board stock. In operation, it slides on top of the stator as shown, covering 17 of the stator plates with a small air gap. The outside plates of the slider are driven by a 3 V p-p 50 kHz CMOS signal at 0° and at 180°. One section of slider plates picks up a sine (or, more accurately, a rounded triangle) signal. The other section, the cosine section, is displaced 0.025 in. The stator plates underneath the sine plates on the slider will then, by capacitive coupling to the slider sine plates, assume a signal which varies as approximately sine (position), that is, a square wave at 50 kHz is picked up whose amplitude describes a sine waveform as the slider is moved, with one cycle for each 0.1 in of travel. The stator plates underneath the cosine slider plates will assume a cosine (position) signal. The center section of the slider is composed of two pickup plates, one of which picks up, by capacitive coupling, the signal of the sine stator plates, while the other picks up the cosine signal. The pickup plates feed high-impedance guarded amplifiers. The block diagram of the 723 is a typical multiplate system, as shown in Figure 18.3.

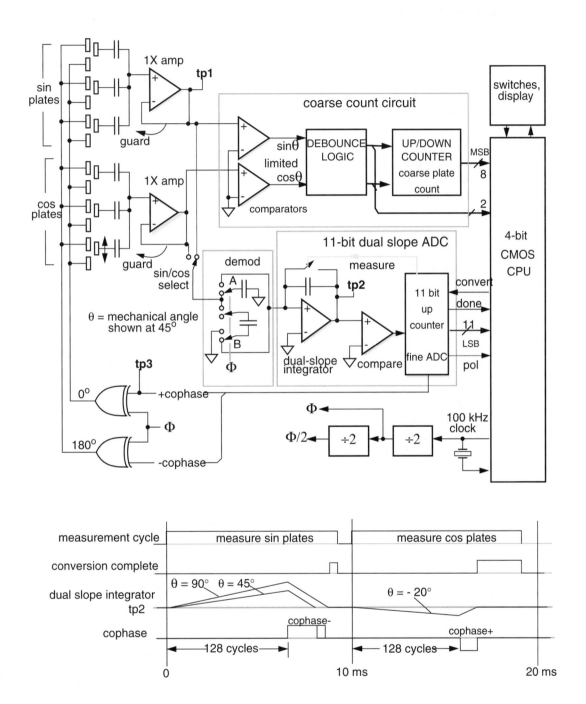

Figure 18.3   Starrett 723 block diagram

18.2   CIRCUIT DESCRIPTION

1x amplifier

Six of the sine plates are shown in the block diagram, coupling through the small 0.85 pF capacitance of the pickup plates to the input amplifier. A high-impedance 1× CMOS amplifier is used, with its output at $tp1$ fed back to guard the pickup. This guard totally surrounds the pickup plate and its connection to the input amplifier, and is connected to the two pins on the amplifier adjacent to the input pin to null out almost all of the capacitance to ground which would otherwise severely attenuate the signal. A small unguarded capacitance on the chip causes about a 35% signal amplitude loss which is compensated by the demodulation circuit.

The amplifier output signal at $tp1$ shows the 50 kHz modulation at a mechanical angle θ of 45° (Figure 18.4).

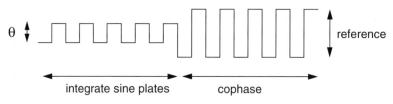

Figure 18.4   Amplifier output signal

The first segment at an amplitude θ changes with mechanical position in the range of 2 V p-p at 50 kHz 0° to 2 V p-p at 180°. The second segment at a reference amplitude results from driving both 0° and 180° sine plates with an identical (either 0° or 180°) signal when the cophase command is issued so that a full-scale reference output is produced. This is similar to the 90° modulation discussed in Chapter 10, except the 90° signal is further modulated by a higher frequency (50 kHz) carrier for this application.

Coarse plate count

The coarse plate count is implemented by squaring the sin and cos input signals with two comparators and determining phase with a D-type flip-flop clocked with 100 kHz. The flip-flop output will be a one for mechanical angles between 0° and 180° and a zero for 180° to 360°. After processing by the debounce logic of Section 8.3, the debounced output signals control an up/down counter which keeps track of coarse position.

Demodulator

The sine and cosine plates are alternately selected at about a 100 Hz rate. The chosen plate's signal is sampled (when the clock square wave is a one level) onto a small holding capacitor by demodulation switch A, and applied to the input of the dual-slope integrator when the clock is a zero. Similarly, switch B samples the input signal when the clock is a zero, and applies the sample capacitor with reverse polarity to the input of the dual-slope integrator. This is a switched capacitor implementation of a conventional synchronous push-pull demodulator followed by an integrator.

Dual-slope integrator

The integrator output, as shown in the timing diagram on Figure 18.3, will slowly increase with a small fixed positive mechanical angle θ to a voltage $K\theta$. After 128 clock cycles, a timer will change the drive signals 0° and 180° so they are both at 0°. This produces a mechanical-angle-independent full-slope discharge of the dual-slope integrator which continues until the integrator reaches zero and a comparator terminates the measurement cycle. During this "cophase +" discharge, the ADC counts clock pulses; full-scale voltage $K\theta$ on the integrator takes exactly 128 clock cycles to discharge to zero, and smaller voltages take proportionately less time. The cophase signal, then, has a time duration proportional to the magnitude of the voltage from the sine plates. It feeds the count enable input of an 11 bit counter which is clocked by 100 kHz. Negative mechanical inputs produce a negative-going slope on the integrator, and during cophase time the cophase term is active, resulting in a full-scale charge of the integrator to zero and a negative sign on the ADC output. This type of integrator is used for digital voltmeter applications, and is insensitive to component drift provided only that an accurate timebase is used.

18.2.1 CMOS ASIC

Sensor interfacing is handled by a CMOS semi-custom integrated circuit, or ASIC (Application-specific Semi-custom Integrated Circuit). This circuit, with a chip size of 3 × 4 mm, uses p- and n-channel MOSFETs from a standard array part with connections done by final-level metallization. This array also provides MOS capacitors in small values and diffused resistors. Macrocells have been characterized by the vendor for amplifiers and logic functions.

The ASIC's advantages are very low device interelectrode capacitance and low power consumption. Stable, small-value MOS capacitors and relatively unstable resistors make it more suited for switched capacitance circuits than conventional resistor-transistor circuits.

18.3   PROBLEMS

Several problems were encountered during design verification.

Wandering baseline

The coarse count comparators were originally designed with a simple circuit, shown above. In operation, the coarse plate count often made errors which were traced to a wandering baseline of the 1× amplifier output signal (Figure 18.5). With a very high impedance resistor (or, in this case, back-to-back high impedance FETs), small static charges could swing the amplifier DC level by 500–1000 mV. This effect was repaired by a change in logic: instead of comparing the signal to zero, one half-cycle of the signal was stored on a small MOS capacitor and compared to the second half of the signal.

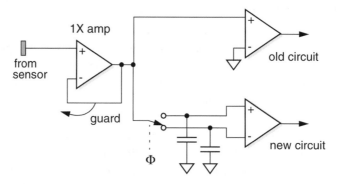

Figure 18.5   Starrett 723, fix for wandering baseline

Static charge

Static charge was a major problem. Static charge effects were not seen early in the design, which was started in the spring, as spring and summer seasons in New England tend to be humid and humidity masks static charge. In the winter, however, the humidity drops and static charge can become a problem. In the 723, the stator plates were covered with a thin layer of glass-epoxy laminate for protection, and this layer could easily pick up a charge after being rubbed on cloth. This charge could discharge through the small air gap to the slider plates and cause an error in the coarse plate count. As the caliper has no way of sensing absolute position, this coarse count error becomes a 0.050 in dimensional error.

Luckily, the effect is short and occurs during only one clock pulse. The repair was to add logic to detect a single comparator output bit which was of the opposite polarity of each of its neighbors, and reset it to its two neighbors' value (Figure 18.6).

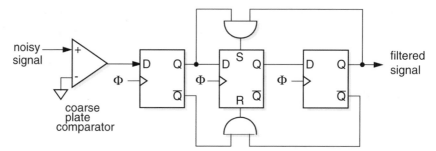

Figure 18.6   Starrett 723, fix for static charge

Parametric capacitance variation

The CMOS ASIC uses p-n diodes from its inputs to the power rails to protect against electrostatic discharge damage; these diodes have the typical variation of capacitance with reverse bias which is caused by the depletion area width becoming modulated by input voltage. The effect on the circuit is a capacitance at the pickup amplifier input which is modulated by the signal. This effect is a fraction of a pF, but as the pickup capacity is also a fraction of a pF, the signal amplitude is modulated in a nonlinear curve (Figure 18.7).

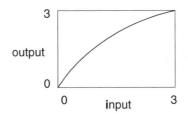

Figure 18.7   Parametric capacitance nonlinearity

For this chip, the p-n diode to the positive rail has the most effect; as the input voltage approaches the positive rail, the capacitance of this diode increases and the signal amplitude compresses.

Nontriangular sine and cosine waveforms

With zero capacitor plate spacing, or with ideal no-fringe-effects capacitors, the interpolation between plates would be ideal triangular waveforms, and the interpolation value could be easily determined by the methods of Section 8.1.1. In practice, the triangle is rounded by fringe effects and the sine and cosine waveforms are not equal value (Figure 18.8).

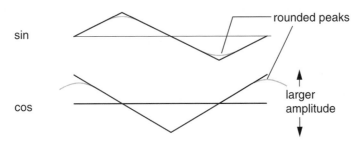

Figure 18.8   Sine and cosine signal inaccuracy

Table lookup

The effects of the shape of the triangular signals and the nonlinearity caused by parametric capacitance variation are both corrected using a lookup table whose entries are empirically determined. As the geometry of the plates and the physical size of the p-n diodes are both stable, the table results apply quite well to different units in production, with no trimming for individual variations needed.

19

Graphic input tablet

Shintron Co., of Concord, MA, manufactured a computer graphic input tablet called Ecricon. This product was developed in 1967, and a U.S. patent was granted in 1971 [Patent 3,591,718 by S. Asano and L.K. Baxter, assigned to Shintron Company, Inc.]. The Ecricon tablet measured the x, y, and z movements of a small capacitively coupled pickup stylus relative to a square-wave-driven resistive sheet which generates an electrostatic field over an 11 in square tablet. Two different methods of producing the orthogonal field lines were investigated, and a novel phase-locked-loop demodulator was used with ratiometric response and good performance at large stylus-to-tablet separation.

19.1 SPECIFICATIONS

Resolution . . . . 1024×1024
Accuracy 1%
Sample rate . . . . 100 xy samples/s
Output format . . . . 10 bit parallel digital
Maximum paper thickness . . . . . . 1/2 in

19.2 GENERATING THE ELECTRIC FIELD

19.2.1 One dimension

Driving a resistive sheet to measure a *single* axis is simple (Figure 19.1). If metallic electrodes A and B are fed with a 5 V square wave, the resistive sheet will generate a linear AC voltage field just over its surface. A stylus using a small electrode will pick up a signal proportional to the y displacement when moved near the surface of the sheet. With circuits which measure signal amplitude, the z position can also be measured, although not linearly.

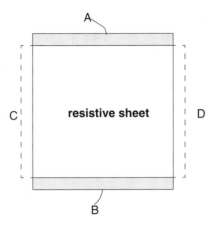

Figure 19.1   Driving a resistive sheet

19.2.2   Two dimensions, resistive sheet

Extending this method to two dimensions is not as simple. If metallic electrodes C and D are added at the dotted lines, the original electrodes A and B will be shorted at the corners, or at minimum a very nonlinear field will be produced. One way around this trap is to use a large number of current generators in parallel to drive the edges of the resistive sheet, and in fact a computer graphics input tablet was built this way by Sylvania, Inc. A simpler method is to drive the corners instead of the edges, and to use a medium resistivity and a high resistivity material to produce an orthogonal, linear field (Figure 19.2).

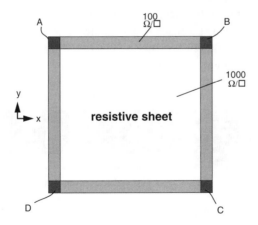

Figure 19.2   Resistive tablet with x and y outputs

To measure position in the y axis, electrodes A and B are connected together and driven with 100 kHz 0° and electrodes C and D are connected together and driven with 100 kHz at 180°. Then as the stylus is moved in the y axis, the electrical angle will change from 0 to 180°, but displacement in x will not affect the signal. Position in the x axis is then determined by driving electrodes A and D together, and B and C together.

Nonlinearity correction

The field produced by this tablet will be linear and orthogonal, excluding small fringe effects, if the ratio of the resistivities is large. In practice, resistivities may be difficult to obtain with high ratios, and a geometric compensation is needed to correct the resulting nonlinearity (Figure 19.3). The curvature is 10% of the tablet size for the 10:1 resistivity ratio shown.

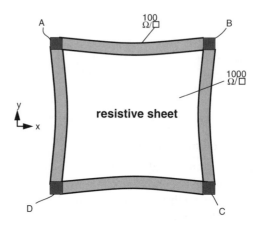

Figure 19.3   Resistive tablet with linearity corrected

With a stored linearity correction table, extreme nonlinearity can be handled; the low resistance strips can be dispensed with, for example, and the resulting over-50% non-linearity compensated by a table lookup operation.

An intermediate approach to generating a linear orthogonal field is to deposit stripes of conductor on 1000 $\Omega/\square$ resistivity material to form the 100 $\Omega/\square$ areas with only one resistive ink printing step (Figure 19.4).

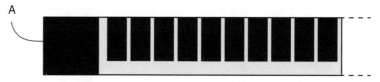

Figure 19.4   Electrodes to reduce 1000 $\Omega/\square$ to 100 $\Omega/\square$

The stripes are about 10× wider than the gap, and a small strip of unstriped area is left so that the field loses its lumpy character by the time it reaches the active area.

Resistive film

Some materials which can be used for resistive films are shown in Table 8.2. Vacuum-deposited thin films are expensive for large substrates, so screened thick film layers are preferred. Pyrolytic deposition of SnO_2, tin oxide, or $InSnO_2$, indium tin oxide, is used when a transparent conductor is needed.

19.2.3 Potentiometer wire

Due to variation of resistivity, the accuracy which can be obtained with the use of resistive film in large sizes may be inadequate, perhaps 5–15% nonlinearity and 50% absolute accuracy. The absolute accuracy is not important in this application, but the nonlinearity can be excessive for many uses. One fix is to store a table of correction values. Another method is to use potentiometer wire rather than film, as fine nichrome wire with excellent resistance accuracy is available for wirewound resistors and potentiometers. A prototype tablet was built (Figure 19.5) with crossed noncontacting grids of 0.005 in diameter nichrome wire, and demonstrated excellent accuracy in the range of 0.1–0.2%.

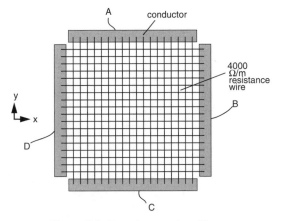

Figure 19.5   Potentiometer wire tablet

Two isolated sheet resistors are formed, one by conductors *A* and *C* and the interconnecting resistance wire, and the other by conductors *B* and *D*. In operation, first a square wave is impressed across *A* and *C* while *B* and *D* are grounded, and then *B* and *D* are driven while *A* and *C* are grounded. The grounded mesh will act as a shield to the driven mesh and reduce the total available field slightly, but as a ratiometric detection scheme is used, the effect is insignificant.

19.3   PICKUP

Stylus

The stylus pickup is a guarded, coaxial construction (Figure 19.6). The stylus used for Ecricon is a standard ballpoint pen with 2 mm of its point exposed for writing and for sensing. Its performance is good up to 10 cm stylus-to-tablet spacing, at which distance its capacitance to the resistive sheet is approximately 0.05 pF and its signal amplitude is reduced to 10% of the amplitude at the surface due to fringe effects and unguarded stray capacitance. This level of performance requires amplifier input capacitance to be less than 0.1 pF; with this low value of amplifier input capacitance the signal is attenuated by 3×, but the ratiometric detection method makes this less important.

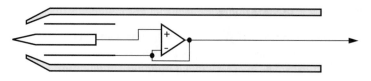

Figure 19.6   Stylus

Pickup circuit

The pickup uses discrete transistors, and both guarding and neutralization to minimize input capacitance (Figure 19.7).

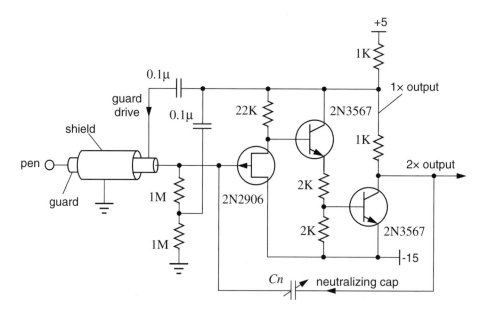

Figure 19.7   Pickup schematic

With this circuit, the guard handles most of the input capacitance cancellation, with a small residue caused by the finite amplifier gain and the gate-to-drain capacitance of the FET. The neutralizing capacitor, with a value of 0–0.1 pF, is adjusted to null out these factors and produce a nominal input capacitance of zero. If Cn is adjusted for a slightly negative input capacitance the circuit is stable with high signal level, but with decreasing signal level, as the stylus is moved away from the tablet, the circuit becomes unstable and oscillates at a few kHz. Luckily, parasitic capacitance is reasonably repetitive and stable, so the neutralizing adjustment is not sensitive to environmental factors once adjusted.

19.3.1   Reverse sensing

If the tablet is floated and touched with a grounded stylus or a fingertip, the signal can be picked off the tablet rather than the stylus. As a large parasitic capacitance is typical with this connection, a virtual ground or feedback-type amplifier is used. This technique is used for CRT touchscreens, with a transparent conductor providing the resistive sheet; for large tablet sizes, a tracking algorithm is more resistant to noise as only the field near the touch is of concern.

19.4   EXCITATION

The signal strength varies over a large range as the stylus is moved away from the tablet, so an amplitude-independent ratiometric demodulator must be used, and the design of the drive voltages must include a reference signal. The 90° drive shown previously (Figure 1.36) is used. One additional modulation is performed, however: the 0 and 90° signals are exclusive-OR'd with 102 kHz to provide a higher frequency carrier which is easier to couple to the pickup and provides greater immunity to power-frequency noise coupling.

19.5   DEMODULATION

The demodulator is a novel phase-locked-loop circuit which handles three tasks:

• Amplitude-independent ratiometric demodulation
• Variable bandwidth reduction
• Analog-to-digital conversion

The demodulator circuit is unusual (Figure 19.8).

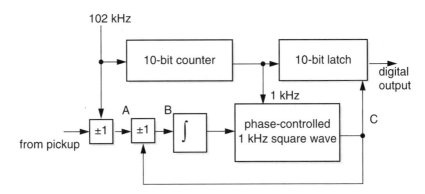

Figure 19.8   Demodulator block diagram

The 102 kHz modulation is demodulated by the first ±1 block, and the signal output of this block is similar to the bottom trace of Figure 1.36. This is then multiplied by the variable-phase square wave output in the second ±1 block and integrated. The integrator time constant is set so that the peak-to-peak AC at its output is small, say 1 V, to avoid saturation effects. The integrator then feeds a voltage-variable phase shift circuit, and the loop is closed back to the second multiplier. This circuit phase locks the square wave so that its phase shift is directly proportional to the ratio of the variable signal to the peak signal.

The operation is explained by the waveform with the 102 kHz carrier subtracted (Figure 19.9). The output at point A is the 500 Hz signal, the linear combination of 0° and 90° square waves after demodulating the 102 kHz carrier with the first ±1 stage. As the stylus moves full range, the variable sector moves linearly from the negative reference voltage to the positive reference voltage. This produces a DC offset at the integrator output which changes the phase of the square wave in a direction to null the offset, so that the three areas shown for point B have the relationship $S = R1 - R2$. As the input signal changes through full scale, the output phase changes linearly through 90°. The feedback will create an undamped second-order response which will overshoot unless the integrator is lossy. The loss can be adjusted for an appropriate damped response.

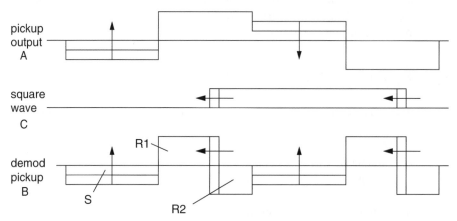

Figure 19.9   Phase-lock waveforms

This demodulator is amplitude-insensitive; it integrates a full cycle so impulse noise is rejected, and it has a variable bandwidth to reject noise. The signal bandwidth is a function of the integrator gain times the input amplitude times the phase-angle-per-volt response of the phase-controlled square wave generator, so as amplitude decreases with increasing stylus height, the gain and bandwidth automatically decrease to reject noise. Another advantage is the precision and simplicity of the digital output.

Although implementation is shown for clarity with two successive ±1 stages, the demodulator can be simplified by combining the two stages into a single ±1 stage with the control signal created by the exclusive-OR of the 102 kHz clock and the phase-modulated square wave.

20
Camera positioner

PictureTel needed a fast and precise camera positioner for video conference applications, and built a camera controllable in zoom, tilt, and pan axes using servoed DC motors. PowerCam (Figure 20.1) moves a 0.5 kg color video camera through a pan (horizontal) axis rotation of 180° in 1 s, and a tilt rotation of 45° in 1/2 s. The motion control design uses a capacitive sensor for smooth and accurate position control in two axes, and illustrates the use of capacitive sensing for high-performance, low-cost closed loop servo systems.

Figure 20.1 PowerCam (patent pending)

20.1   SPECIFICATIONS

PowerCam's lens can be controlled in zoom from 7.5° to 75° field of view. At the narrow 7.5° field of view, the camera must be moved very smoothly to avoid perceptible jerkiness in the video; conservatively, jitter of a quarter of a pixel can be seen with very slow motion. With its 768-horizontal-pixel imager, a single pixel at maximum zoom is 7.5/768 = 0.01° of pan motion. The resolution requirement is more than $4 \times 270/0.01 = 108,000:1$, or 17 bits. If a digital system is used, perhaps 19 bits of resolution would be needed to avoid limit cycle behavior at the LSB level.

Motion specifications

Pointing accuracy	1.5°
Pointing repeatability	0.25°
Pan axis range	270°
Pan axis speed	180° rotation in 1 s
Tilt axis range	45°
Power	12 V, 1 A

Other design parameters are low cost and low acoustic noise.

20.2   MOTOR

Motor selection began with an analysis of required torque. The pan axis is shown, as it is the worst case for motor torque. The camera polar inertia can be approximated by this solid shape (Figure 20.1a).

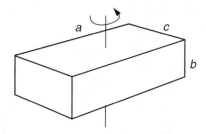

Figure 20.1a   Camera inertia

With the 0.5 kg mass evenly distributed, the polar inertia with a = 13 cm and c = 9 cm is calculated as

$$J = mk^2 = 0.5\left(\frac{a^2 + c^2}{12}\right) = 0.001 \ \text{kg-m}^2 \qquad 20.1$$

Assuming the acceleration profile has no coast, the motion looks like Figure 20.2.

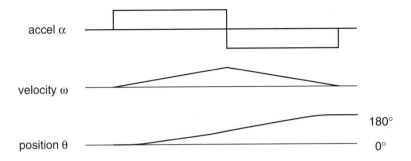

Figure 20.2   Camera motion profile

The minimum acceleration α to meet the 180°/s speed specification is calculated from

$$\theta = \frac{1}{2}\alpha t^2 \tag{20.2}$$

so that

$$\alpha = \frac{2\theta}{t^2} = \frac{\pi}{0.5^2} = 12.56 \quad \text{rad/s}^2 \tag{20.3}$$

And the torque to accelerate the load is

$$T = J\alpha = 0.01335 \quad \text{N-m} \tag{20.4}$$

or 1335 g-cm. Assuming a 75% transmission efficiency, a relatively low 50:1 gear ratio to keep motor noise low and a safety factor of 2, peak motor shaft torque, ignoring motor inertia, is 71 g-cm at 1500 rpm.

   Of the many types of motors which could be considered, four were closely compared (Table 20.1).

Table 20.1 Motor comparison

	DC brush	DC brushless	Stepper PM	Stepper hybrid
Cost, 10K qty.	$1.25	$6.00	$5.00	$18
Efficiency	0.6	0.65	0.25	0.35
Life	>1500 h	>10,000 h	>10,000 h	>10,000 h
Friction	2–5 g-cm	lowest friction; depends on bearings		
Size, length × diam.	3 × 2 cm	3 × 2 cm	4 × 3	5 × 4
Max. speed	8000 rpm	8000 rpm	1500 steps/s	5000 steps/s
Drive	power amp	Hall-effect commutate	microstepping drive is needed for smoothness	

20.2.1 DC brush motor

DC brush motors have high efficiency and very low cost, and are the easiest to drive. DC brush motors designed for servo use have low inertia rotors, five or more commutator segments for smooth torque, a large diameter shaft so shaft resonance does not affect servo stability, and they are well characterized for servo use. Unfortunately, the price reflects this additional sophistication.

Small mass-production DC brush motors are available in the $1–5.00 price range, from vendors like Mabuchi, Canon, Chinon, Pittman, and Kollmorgan with three-segment commutators, sleeve bearings, and smaller (1–1.5 mm) shaft diameters.

EMI radiation can be a problem with DC brush motors, as sparking at the commutator generates broadband noise. A ceramic capacitor close to the brushes will help. Some models, such as Mabuchi's RF series, handle EMI with an internal varistor.

Torque ripple

Two motor design details affect smooth torque delivery. One is the higher torque ripple with a three-segment commutator (Figure 20.3).

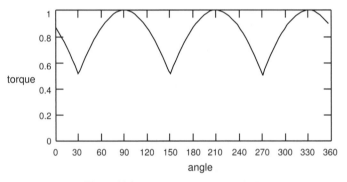

Figure 20.3   Three-segment torque ripple

The three-segment commutator has a 50% torque ripple. If the motor is enclosed in a high-gain servo loop, this may not be a problem. With five segments, the torque ripple improves to 82% (Figure 20.4).

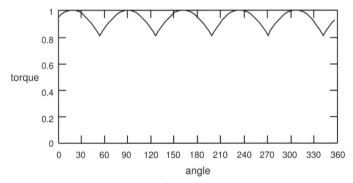

Figure 20.4   Five-segment torque ripple

Several three- and five-pole motors were tested for torque vs. shaft angle performance. Torque curves were similar to above predictions, except magnetic gap variations produced a 5–20% variation of peak torque as a function of angle.

Neutral commutation

Commutation in a brush-type DC motor at slow speed should switch the armature current from one coil to the next just as the coils are equally positioned around the magnetic pole, so the active coil is switched out just as the next coil is in position to produce an equal torque. This is called neutral commutation, and a motor designed for low speed or bidirectional operation will have neutral commutation. At high speed, the inductance of the armature slows the rate of armature current change, and the brushes would need to be advanced slightly to avoid a torque transient and arcing at the switch point. Motors with this feature have a preferred speed and direction and should not be used for servo actuators.

Commutation transient

Another potential problem with DC motors is the momentary torque transient which can be generated at the commutation point if the brushes are not correctly positioned, or if the magnetic circuit or the windings are not accurate, or when a brush bridges two commutator segments momentarily. This problem is reduced with five-segment armatures, and with current drive circuits instead of voltage drive.

20.2.2 DC brushless motors

DC brushless motors have excellent efficiency, longer life than brush motors, and are reasonably low in cost. One potential problem for smooth, low-speed positioning servos is the commutation transient. While the relatively slow torque ripple of a three-segment commutator can usually be attenuated by a high-gain servo loop, a quick transient caused by inaccurate commutation cannot.

More detail on DC brushless motors is found in Section 9.4.

20.2.3 Permanent magnet step motors

Permanent magnet, or PM, steppers are reasonably priced, in the $5.00 range for medium quantity, and have 3.75, 7.5, and 15° step angles. A large range of sizes is available, down to about one cm diameter. The smaller sizes are capable of faster stepping, so maximum output power does not scale directly with size, but smaller sizes are limited to the coarser step angles. The step sizes are much too coarse for smooth motion in the current application, but microstepping drives can improve this.

Microstepping replaces the usual square wave drive with sine waves in quadrature, and in theory can produce arbitrarily smooth motion. The difficulty in practice is that the PM stepper is not accurately constructed, and several effects conspire to degrade smooth low speed motion:

1. Quadrature relationship. The nominal 90° relationship of the two windings may be several degrees misaligned.
2. Quadrature amplitude. The two windings may need different current amplitudes for good accuracy.

3. Detent torque. The rotor is detented by the magnetic circuit to each step angle. Increasing the size of the magnetic gap improves microstep width at a cost of efficiency.
4. Hysteresis. Remnant magnetism near winding current zero cross causes small nonlinear effects when current is reversing.
5. Resonance. The magnetic restoring torque and the combination of motor and load inertia form a mechanical resonance, usually in the 200–500 steps/s range. As the motor passes through this resonance, the operation can be rough.

These effects can be considerably ameliorated by careful shaping of the drive signals, and adjusting their phase relationship and amplitude. The ideal drive waveform often becomes a cusp shape, reversing the expected sine shape, to repair detent torque effects. The improvement afforded by these methods is limited, however, as imperfections in the rotor magnetization cause each step's ideal drive current to be slightly different. A practical limit is 8–10 microsteps/step. Other drawbacks with stepper drives for this application are poorer efficiency and relatively noisy operation. Performance is improved by using feedback from output shaft position to the motor driver to compensate for these nonlinearities, but the stepper in this system has little to offer compared to brush or brushless DC motors.

20.2.4 Hybrid step motors

Hybrid steppers use a combination of magnetic hysteresis and a permanent magnet rotor. They feature higher efficiency than PM types, higher speed operation, and smaller step angles to 1.8°, but the price is higher. Low speed smoothness is better than in PM steppers, but decent performance still requires the adjustments above to shape winding current. Drawbacks for this application are large size (1.75 in minimum diameter) and high cost, over $17.

Motor choice: DC

The brush-type DC motor was chosen for this application for its low cost, low power, and simple drive circuits. For more stringent applications which need longer life or higher pointing accuracy, a brushless DC motor or a carefully driven PM stepper would be the choice.

20.2.5 Transmission

A one-pass friction drive was selected for quiet operation, low cost, zero backlash, and freedom from adjustment. A section of the pan axis is shown in Figure 20.5.

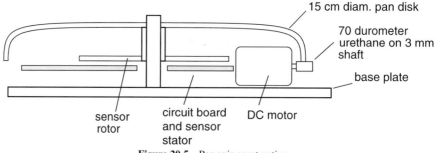

Figure 20.5   Pan axis construction

The gear ratio, 50:1, results in a maximum motor speed of about 3000 rpm. This speed is a little low for maximum motor power output, but helps keep noise low and motor life high. This is not a positive drive, so allover feedback is needed.

20.3   SERVO SYSTEM

The servo uses an output shaft capacitive position sensor and closes a loop with allover feedback from output shaft position to motor drive current. The standard PID (proportional-integral-derivative) loop compensation was chosen after a brief, unproductive romance with fuzzy logic.

The servo loop was built using analog circuits. This makes for a low cost, high performance loop, although it cannot accommodate variations in load inertia of more than a factor of three to four. If the load inertia was not constant, an adaptive digital approach could be used so that the servo parameters could more easily be adapted to match load inertia.

A precision conductive-plastic potentiometer was tried first for position sensing. Its output noise was rated at 0.1% rms which seemed adequate, but when tuning the servo the derivative gain Kd needed to be high for good slow-speed smoothness. This high gain amplified the potentiometer noise enough to swing the power amplifier output rail-to-rail, limiting gain and efficiency.

Position sensor choice: capacitive encoder

The final position sensor used was a capacitive type (Figure 20.6) with 1.6" outside diameter; its noise is less than 0.005% rms. The dual V geometry shown previously (Figure 7.8) was used, but as a rotary pattern rather than linear, with the pickup plate stationary. The wiring to the rotor is free in this design, as travel is limited to 270° and a wire cable is needed for other camera signals. The rotor drive voltages are at 5 V p-p and are low impedance so no special shielding is needed. The fixed pickup location means that this very high impedance point is easy to shield and guard with PC etch, and its connection to the amplifier can be minimum length.

Both rotor and pickup are fabricated from FR-4 0.062 in PC board stock, with bare nickel plating on the electrodes rather than solder mask to avoid triboelectric effects.

The circuit chosen is a 0–180° drive, with feedback-type amplifier for low spacing sensitivity, with an empirically determined positive feedback to make sensor amplitude variation with spacing nominally zero. Because an off-the-shelf operational amplifier has too much input-to-output capacity, which results in low output voltage and excessive spacing dependence, a discrete FET transistor input preamp was added (Figure 10.30). Demodulation uses a 74HC4053 CMOS switch and a lowpass filter with a 580 Hz corner frequency so that its phase shift at the servo zero cross frequency (about 100 Hz) is low.

With an air gap of 0.5 mm, the demodulated output range is between 0.4 and 4.6 V instead of the theoretically expected 0 to 5 V, representing a stray amplifier input-to-output capacitance of less than 0.16 pF.

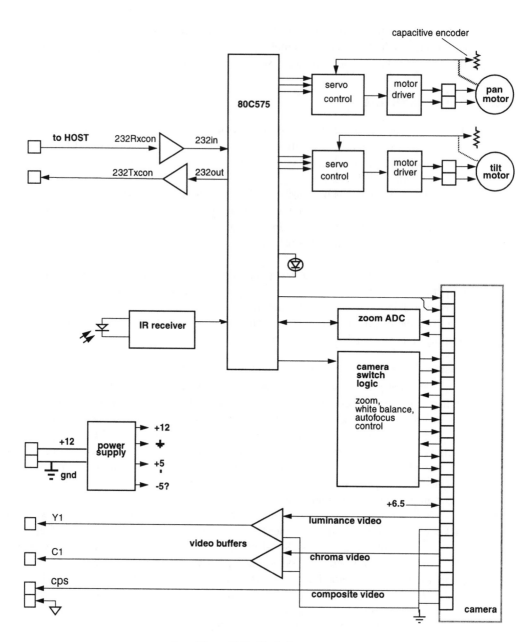

Figure 20.6   Block diagram

21
Digital level

SmartTool Technologies of San Jose, CA (a division of Macklanburg-Duncan) specializes in smart tools, tools that make construction jobs easier. An example is the SmartLevel® (Figure 21.1) which looks like the ordinary spirit level with a digital display. But where the spirit level is capable of measuring only inclinations of 0° and 90°, SmartLevel handles a full 360°. RS-232 serial output is available in some models, and a hold button simplifies hard-to-get-at measurements. Display can be angle in degrees or rise/run or angle in percent, and the display reads right-side-up even when the level is inverted, courtesy of the level's gravity sensor. Smaller models are available, 6" long, for machinists and model makers. Resolution can be as good as 0.01° and accuracy in some models is 0.05° at small angles. An OEM module with just sensor and electronics occupies a 1.3 × 2.6 × 0.6" volume, with the pillbox-size capacitive sensor occupying about half the space.

The SmartLevel design shows the evolution of the product from a simple concept with minimum performance to a refined design with excellent specifications. More than 500,000 SmartLevels have been sold.

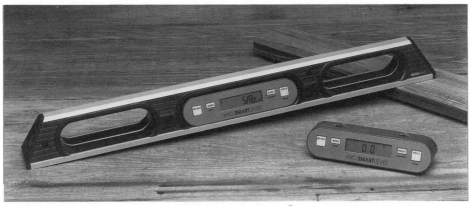

Figure 21.1 SmartLevel®

21.1   SPECIFICATIONS

The product specifications for one version of the sensor module are:

Range	360°
Resolution	
serial port.	0.00549°
PWM	4096 steps
Accuracy	
at level and plumb	± 0.1°
at other angles	± 0.2°
Calibration points	128
Temperature coefficient	
at level or plumb, ± 1°	0.006°/°C
at other angles	0.012°/°C
Repeatability	± 1°
Hysteresis	0.1°
Noise	± 0.02°
Settling time	3 s typ; 5 s max
Angle output rate.	533 ms
Size	6.6 × 3.3 ×1.5 cm
Power	5 V DC, 2 mA
Temperature range	−5 to +50°C operating
Price	$85.00 (1)

21.2   SENSOR

Simple sensor

A simple level sensor, or inclinometer, can be built with rotating air-spaced capacitor plates (Figure 21.2).

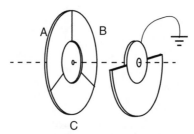

Figure 21.2   Level sensor plates

The disk on the left is a metal plate fixed to the body of the level and divided into three equal 120° segments. A circuit measures the capacity of each plate to ground. The semi-disk on the right is a rotating metal plate on a good bearing, so it tends to hang as shown with gravity. The capacitance of the three plates to ground, normalized to equal plus and minus excursion, changes with tilt angle (Figure 21.3).

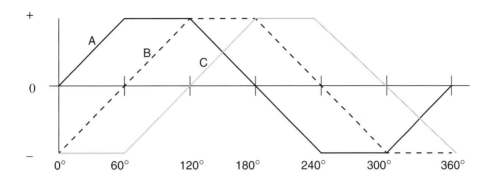

Figure 21.3   Capacitance variation vs. rotation angle

The position can be coarsely estimated to the nearest 60° segment by observing that the signs of *ABC* in the first 60° of rotation are + - -, and that each 60° segment has a unique order of signs. Then the fine position can be calculated by choosing the plate with the lowest voltage (hence changing the fastest) and performing a simple conversion or table lookup to fine pitch angle. Note that the preferred ratiometric techniques shown in Chapter 8 could also be applied, as the three-phase angle measurement is equivalent to the sine-cosine measurement and coarse plate count. Also, the multiplate techniques of Chapter 8 could be applied to trade analog resolution for digital resolution.

Problems with the three-electrode metal-disk method include:

- Hysteresis caused by friction in the bearing
- Spacing variation causes erroneous pitch angle reading
- Drift due to absolute rather than ratiometric measurement of capacitance

Liquid filled sensor

The first improvement, replacing the metal rotor with a partially conductive liquid, has some important advantages. The liquid has insignificant friction, and the air gap alignment problem disappears as the air gap is replaced with a thin, stable dielectric coating on the metal segment plate. The liquid is poured into a flat cylindrical chamber, and with the chamber half full the liquid acts nearly the same as the semicircle of metal, with the periphery of the chamber grounded to ground the liquid. But a problem with the liquid approach is that while the free surface of the liquid tilts correctly with the desired pitch-angle direction of measurement, it also tilts slightly in the orthogonal axis when the level is rolled slightly, producing an unwanted response to roll.

The roll angle sensitivity is improved by placing similar electrodes on opposing faces of the liquid capsule (Figure 21.4). Now, the capacity to ground of pairs of electrodes, *A + D*, *B + E*, and *C + F*, is measured. The benefit is that as the liquid level tilts up on, say, the left side, increasing the capacitance of *A*, *B*, and *C* to ground, the liquid level tilts down on the right side, causing a compensating decrease in the capacitance of *D*, *E*, and *F* to ground, so *A + D*, e.g., is unaffected. See Figure 21.5.

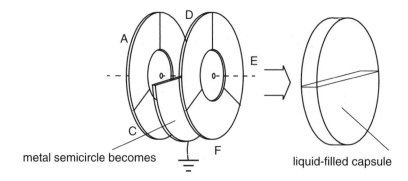

metal semicircle becomes

liquid-filled capsule

Figure 21.4   Liquid-filled level sensor

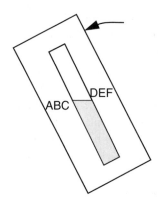

Figure 21.5   Sensor tilted in roll axis

Optimizing the geometry of the sensor plates is also important to resist roll; pie-shaped plates are not optimum. If full pie-shaped plates are used, the increase in capacitance on the *ABC* side is not exactly compensated by the decrease on the *DEF* side if the plates are in some orientations (Figure 21.6). This error, which would amount to about 2°, can be eliminated by use of donut-shaped plates (Figure 21.7).

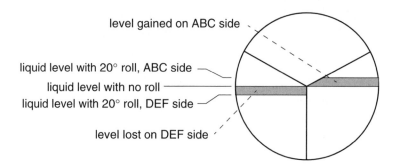

Figure 21.6   Pie-shaped plates produce roll couple

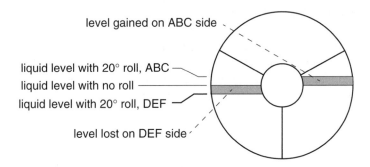

level gained on ABC side

liquid level with 20° roll, ABC

liquid level with no roll

liquid level with 20° roll, DEF

level lost on DEF side

Figure 21.7   Donut-shaped plates eliminate roll couple

Excessive roll will still produce a roll couple, but the addition of two extra electrodes in the hub provides a method of detecting excess roll and flashing the display to warn the user of a potentially inaccurate reading.

21.2.1 Sensor construction

The sensor is 3.2 cm in diameter, and is built using printed circuit board electrodes and an aluminum ring (Figure 21.8). The materials used are carefully chosen. An inert nonconducting layer protects the PC board electrodes and acts as the dielectric for the variable-area capacitor formed by electrode and liquid. The thickness uniformity of the dielectric layer is critical for good linearity.

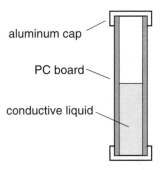

aluminum cap

PC board

conductive liquid

Figure 21.8   Sensor cross section

The liquid needs to be slightly viscous so the liquid-air interface is stable, but not so viscous to add to settling time, and it needs to be inert when in contact with aluminum and have a reasonable surface tension to control the liquid-air interface. A combination of an alkane and a ketone were chosen to optimize these properties. A dielectric, nonconducting fluid could also be used, or a resistive fluid could be measured by resistive rather than capacitive sensing, but a manufacturing tolerance problem would be found in keeping the gap very accurate. With the chosen technique, the dielectric layer needs to be accurate, but the gap can vary.

21.3   CIRCUIT DESCRIPTION

The circuit needs to measure six different capacitors to ground, calculate the angle, apply various calibration factors and display the result. A single 555-type *RC* oscillator is multiplexed with an analog switch to the six electrodes to guarantee stable capacitance measurements that track over temperature (Figure 21.9).

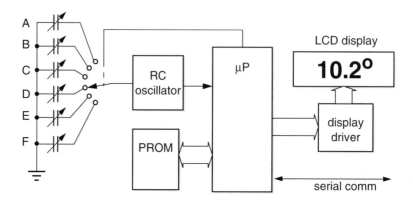

Figure 21.9   Circuit diagram

The PROM is an electrically reprogrammable type which is used to store calibration data, both from a production calibration sweep and from field recalibration, and an LCD display is used to conserve power.

21.4   CALIBRATION

The accuracy specifications of the instrument, 0.1–0.2°, represent 1/1200 of a 120° sector. This is a very difficult target for a single analog measurement, and the best technique to achieve this performance is to calibrate each sensor and store the results in the PROM. Two different types of calibration are needed.

Linearity

The level is rotated slowly through 360° as the last step in production, and automatic test equipment detects the sensor error and stores it in the PROM. There are 128 points of calibration used, and values are interpolated between these points. This calibration compensates for electrode shape error, dielectric thickness error, fill level error, and several other perturbations.

Zero set

During use, another calibration can be performed. The user sets the level on a flat surface and presses the CAL button, reverses the level end-for-end and presses the CAL button again. The microcomputer then averages the readings and can determine the zero offset even if the flat surface is not completely level.

22
References

Accumeasure System 1000. Dec. 1992. MTI Instruments catalog.

AccuRange 3000 Catalog, Acuity Research Incorporated.

Ajluni, C. Dec. 5, 1994. Interoperability: The latest buzzword in sensors. *Electronic Design* 59–68.

————. Oct. 1994. Pressure sensors strive to stay on top. *Electronic Design* 68–74.

Allen, P.E., and E. Sánchez-Sinencio. 1984. *Switched Capacitor Circuits.* New York: Van Nostrand Reinhold.

Artyomov, V.M. 1991. Silicon capacitive pressure transducer with increased modulation depth, *Sensors and Actuators* A28: 223–230.

Bellanger, M. 1984. *Digital Processing of Signals.* New York: John Wiley & Sons.

Berger, W. Mar. 1982. Interfacing digital circuitry with plastic rotary encoders, *Electronics* 128–130.

Berube, R.H. 1993. *Electronic Devices and Circuits Using MICRO-CAP III,* Merrill.

Bibl, A. June 30, 1981. Shielding circuits in plastic enclosures, *Electronic Products.*

Bindicator Level Sensors. 1991. catalog.

Blickley, G.J. Aug. 1993. Sensor performance improved by electronics materials, *Control Engineering* 49–50.

Bonse, M.H.W., C. Mul, and J.W. Spronck. 1995. Finite-element modelling as a tool for designing capacitive position sensors, *Sensors and Actuators* A46–A47: 266–269.

Bonse, M.H.W., F. Zhu, and J.W. Spronck. 1994. A new two-dimensional capacitive position transducer, *Sensors and Actuators* A41–A42: 29–32.

Booton, R.C., Jr. 1992. *Computational Methods for Electromagnetics and Microwaves.* New York: John Wiley & Sons.

Bouwstra, S. 1990. Resonating microbridge mass flow, *Sensors and Actuators* A21–A23: 332–335.

Burger, R.M. "Film Materials and Techniques," Ch. 21 in *Materials Science and Technology*, A.E. Javitz, Ed. New York: Hayden Book Co., Inc. 1972.

Chamberlain, G. Dec. 1994. DN 100, *Design News.*

Charles, R.E. Oct. 1966. A capacitance bridge assembly for dielectric measurements from 1 Hz to 40 MHz, *Technical Report 201*, Laboratory for Insulation Research, MIT.

Cho, S.T. 1993. A high-performance microflowmeter with built-in self test, *Sensors and Actuators* A36: 47–56.

Colclaser, R.A. 1980. *Micro Electronics Processing and Device Design.* New York: John Wiley & Sons.

Core, T.A. Oct. 1993. Advanced device packing, *Solid State Technology.*

———. Oct. 1993. Fabrication technology for an integrated surface-micromachined sensor, *Solid State Technology.*

Data Instruments Linear and Pressure Transducers Catalog.

Data Instruments, Pressure and Displacement Transducers, July 1994.

Data Instruments Short Form Catalog.

Davis, A.M. Nov. 5, 1979. Switched-capacitor techniques implement effective IC filters, *EDN* 103–108.

de Bruin, D.W. Sept. 1992. Two-chip smart accelerometer, *IC Sensors.*

de Jong, G.W., G.C.M. Meijer, K. van der Lingen, J.W. Spronck, A.M.M. Aalsma, and D.A.J.M. Bertels. 1994. A smart capacitive absolute angular-position sensor, *Sensors and Actuators* A41–A42: 212–216.

Donn, E.S. Nov. 1971. Measuring digital error rate with pseudorandom signals, *Telecommunications.*

Dunn, W. Automotive sensor applications, abstract, Semiconductor Research and Development Laboratories.

East, B.B. May 1964. Dielectric parameters and equivalent circuits, *Technical Report 189*, Laboratory for Insulation Research, MIT.

Eichenberger, C. Feb. 1990. On charge injection in analog MOS switches and dummy switch compensation techniques, *IEEE Transactions on Circuits and Systems* 37(2).

Eshbach, O.W., Ed. 1936. *Handbook of Engineering Fundamentals.* New York: John Wiley & Sons.

Fink, D.G. 1987. *Standard Handbook for Electrical Engineers, 12th ed.* New York: McGraw-Hill.

Fish, P.J. 1994. *Electronic Noise and Low Noise Design.* New York: McGraw-Hill.

Fletcher, P. Jan. 23, 1995. French laboratory develops a process for commercialization of micromachined sensors, *Electronic Design* 36–38.

Franklin, R.C. et al. July 4, 1978. Electronic wall stud sensor, U.S. Patent 4,099,118.

Giant magnetoresistance technology paves the way for magnetic field sensing. Nov. 21, 1994. *Electronic Design* 48.

Giesler, T. 1993. Electrostatic excitation and capacitive detection of flexural plate-waves, *Sensors and Actuators* A36: 113–119.

Ginsberg, G.L. Aug. 1985. Chip and wire technology: The ultimate in surface mounting, *Electronic Packaging & Production* 78–83.

Hamamatsu catalog, *Optical Displacement Sensors*, H3065 Series.

Hardy, J. 1979. *High Frequency Circuit Design.* Reston, VA: Reston Publishing Company (a Prentice-Hall Company).

Harper, C.A., and R. Sampson. 1970. *Electronic Materials & Processes Handbook.* New York: McGraw-Hill.

Haus, H.A. 1989. *Electromagnetic Fields and Energy.* Englewood Cliffs, NJ: Prentice-Hall.

Hayt, W.H., Jr. 1958. *Engineering Electromagnetics.* New York: McGraw-Hill.

Heerens, W.Chr. 1986. Application of capacitance techniques in sensor design, *Journal of Physics E: Scientific Instruments* 19: 897–906. Bristol, England: Institute of Physics Publishing.

Hodgman, C.D., Ed. 1954. *Handbook of Chemistry and Physics.* Cleveland, OH: Chemical Rubber Publishing Co.

Hoffman, R.D.R. July/Aug. 1994. Touch screens for portables, *OEM Magazine*, 66–74.

Holman, A.E. 1993. Using capacitive sensors for *in situ* calibration of displacements in a piezo-driven translation stage of an STM, *Sensors and Actuators* A36: 37–42.

Hosticka, B.J., and W. Brockherde. The art of analog circuit design in a digital VLSI world, CH2868-8/90/0000-1347, IEEE.

Intema, D.J. 1992. Static and dynamic aspects of an air-gap capacitor, *Sensors and Actuators* A35: 121–128.

Javitz, A.E., Ed. 1972. *Materials Science and Technology for Design Engineers*. New York: Hayden Book Company.

Jonson, C.R. Dec. 5, 1994. TouchPad can model human senses, *Electronic Engineering Times* 44.

Jones, R.V., and J.C.S. Richards. 1973. The design and some applications of sensitive capacitance micrometers, *Journal of Physics E: Scientific Instruments* 6: 589–600.

Jordan, E.C. 1950. *Electromagnetic Waves and Radiating Systems*. Englewood Cliffs, NJ: Prentice-Hall.

Jordan, J.R. 1991. A capacitance ratio to frequency ratio converter using switched-capacitor techniques, *Sensors and Actuators* A29: 133–139.

Keister, G.W. Apr. 1992. Sensors and transducers, *ECN* 21–32.

Kielkowski, R. 1993. *Inside SPICE—Overcoming the Obstacles of Circuit Simulation*. New York: McGraw-Hill.

Kjensmo, A. 1990. A CMOS front-end circuit for a capacitive pressure sensor, *Sensors and Actuators* A21–A23: 102–107.

Kobayashi, H. 1990. A study of a capacitance-type position sensor for automotive use, *Sensors and Actuators* A24: 27–33.

Kosel, P.B. June 1981. Capacitive transducer for accurate displacement control, *IEEE Transactions on Instrumentation and Measurement* IM-30(2).

Krauss, H.L. 1980. *Solid State Radio Engineering*. New York: John Wiley & Sons.

Kuehnel, W. 1994. A surface micromachined silicon accelerometer with on-chip detection circuitry, *Sensors and Actuators* A45: 7–16.

Kung, J.T., H.S. Lee, and R.T. Howe. 1988. A digital readout technique for capacitive sensor applications, *IEEE J. Solid State Circuits* 23: 972.

Kung, J.T., and H.S. Lee. 1992. An integrated air-gap capacitor pressure sensor and digital readout with sub-100 attofarad resolution, in *Proc. IEEE/ASME J. Micro Electro Mechanical Systems* 1: 121.

Kuzdrall, J. Sept. 17, 1992. Build an error-free encoder interface, *Electronic Design*.

Lang, T.T. 1987. *Electronics of Measuring Systems*. New York: John Wiley & Sons.

LVDT Product Catalog, Robinson-Halpern Company.

Magnetrol, catalog.

Martin, K. Jan. 7, 1982. New clock feedthrough cancellation technique for analogue MOS switched-capacitor circuits, *Electronics Letters* 18(1): 39–40.

Mazda, F.F., Ed. 1983. *Electronics Engineer's Reference Book*. Oxford, England: Butterworth/Heinemann.

Meares, L.G., and C.E. Hymowitz. 1988. Simulating with Spice, Intusoft, San Pedro, CA.

Microlin, Linear & Rotary Displacement Transducers, catalog.

Mini-Circuits Catalog.

Moore, B. Feb. 4, 1988. Consider the tradeoffs when evaluating linear-semicustom ICs, *EDN* 135–142.

Morrison, R. 1977. *Grounding and Shielding Techniques in Instrumentation, 2nd ed*. New York: John Wiley & Sons.

Motchenbacher, C.D. 1973. *Low-Noise Electronic Design*. New York: John Wiley & Sons.

———. 1993. *Low-Noise Electronic System Design*. New York: John Wiley & Sons.

Motorola, *MPX Series Pressure Sensor Elements Catalog.*

MTI 1000 Fotonic Sensor, catalog.

NRD Products Catalog.

Olney, D. Shock and vibration measurement using variable capacitance, Endevco Technical Paper.

Ott, H.W. 1988. *Noise Reduction Techniques in Electronic Systems*. New York: John Wiley & Sons.

PCB Pressure Sensors Catalog.

Pease, R.A. 1991. *Troubleshooting Analog Circuits*. Boston: Butterworth-Heinemann.

————. May 30, 1994. What's all this Teledeltos stuff, anyhow?, *Electronic Design* 101–102.

Precision MELF SMD Resistors, Series 9B. 1992/1993. Surface Mount Device Catalog, Philips Components, 106.

Precision Positioning, ACU-RITE Catalog, Jamestown, NY.

Puers, R. 1990. A capacitive pressure sensor with low impedance output and active suppression of parasitic effects, *Sensors and Actuators* A21–A23: 108–114.

————. 1993. Capacitive sensors: When and how to use them, *Sensors and Actuators* A37–A38: 93–105.

Reference Data for Radio Engineers, 6th ed. 1968. Indianapolis: Howard W. Sams & Co.

Research on high-temperature dielectric materials. Nov. 1965. *Technical Report No. 1*, Laboratory for Insulation Research, MIT.

Riedel, B. 1993. A surface-micromachined, monolithic accelerometer, *Analog Dialogue* 27(2).

Senese, T. Jan. 16, 1990. N-8000/S-glass for CTE matching to leadless ceramic chip carriers, NELCO laminates, Smart VI Conf.

Sensors, Murata Erie North America, catalog no. S-02-A.

Setra, *Pressure Transducers & Transmitters*, catalog.

Schnatz, F.V. 1992. Smart CMOS capacitive pressure transducer with on-chip calibration capability, *Sensors and Actuators* A34: 77–83.

Shimp, D.A. Jan. 1990. Cyanate ester resins—Chemistry, properties, and applications, Hi-Tek Polymers, Inc.

Soloman, S. 1994. *Sensors and Control Systems in Manufacturing.* New York: McGraw-Hill.

Sony, *General Catalog*, Precision Measuring Equipment.

Stout, M.B. 1950. *Basic Electrical Measurements.* Englewood Cliffs, NJ: Prentice-Hall.

Sunx Sensors. 1994. catalog.

Surface Mount Device Catalog. 1992/1993. Philips Components.

Sze, S.M., Ed. 1994. *Semiconductor Sensors.* New York: John Wiley & Sons.

Takahashi, K. Feb. 1994. Bi-MOSFET amplifier for integration with multimicroelectrode array for extracellular neuronal recording, *IEICE Trans. Fundamentals* E77-A(2).

Temposonics II. Sept. 1994. MTIS Systems Corporation.

Terry, S. Aug. 1994. Tradeoffs in silicon accelerometer design, *Sensors.*

Time, R.W. 1990. A field-focusing capacitance sensor for multiphase flow analysis, *Sensors and Actuators* A21–A23: 115–122.

Turck Sensors Catalog, Minneapolis, MN.

Twaddell, W. Nov. 24, 1983. Standard-cell design inroads spur specialized gate-arrray growth, *EDN* 49–58.

Understanding Accelerometer Technology, Catalog, IC Sensors, Milpitas, CA.

UniMeasure, *Linear Position and Velocity Transducers*, catalog no. 400038A.

Vladimirescu, A. 1994. *The SPICE Book.* New York: John Wiley & Sons.

von Hippel, A. Oct. 1954. Dielectric analysis of biomaterial, *Technical Report 13*, Laboratory for Insulation Research, MIT.

———. 1954. *Dielectric Materials and Applications.* Cambridge, MA: Technology Press of MIT.

Walker, C.S. 1990. *Capacitance, Inductance and Crosstalk Analysis*. Boston: Artech House.

Walsh, R.A. 1990. *Electromechanical Design.* Blue Ridge Summit, PA: TAB Professional and Reference Books.

West, J. Sept./Oct. 1994. Micro-processor based position control, *Motion* 10(5): 28–33.

Westerlund, S. Oct. 1994. Capacitor theory, *IEEE Transactions on Dielectrics and Electrical Insulation* 1(5).

Willingham, S., and K. Martin. 1986. Effective clock-feedthrough reduction in switched capacitor circuits, *Journal of Physics E: Scientific Instruments* 19: 897–906.

Wooley, B.A. 1983. BiCMOS analog circuit techniques, CH2868-8/90/0000-1983, IEEE.

Yun, W., and R.T. Howe. Oct. 1991. Silicon microfabricated accelerometers: A perspective on recent developments, in *Proc. Sensors Expo* 204A-1– 204C-12.

Zimmermann, C. and E. Linder. 1991. Capacitive polysilicon resonator with MOS detection circuit, *Sensors and Actuators* A25–A27: 591–595.

Appendix 1

Capacitive Sensors
in Silicon Technology

Reprinted with permission from R. Puers, "Capacitive sensors: when and how to use them," in *Sensors and Actuators*, A 37–38, pp 93–105, 1993.

Authors	Type	Size (mm^2)	C_0 (pF)
Sander, Knutti and Meindl, A monolithic capacitive pressure sensor with pulse period output, *IEEE Trans. Electron Devices, ED-17* (1980) 927–930	pressure	3 × 3	22
Ko, Bao, and Hong, A high sensitivity integrated circuit capacitive pressure sensor IC, *IEEE Trans. Electron Devices, ED-29* (1982) 48–56	pressure	2 × 4	?
Lee and Wise, A batch-fabricated silicon capacitive pressure transducer with low temperature sensitivity, *IEEE Trans. Electron Devices, ED-29* (1982) 42–48	pressure	4 × 4	12
Ko, Shao, Fung, Shen and Yeh, Capacitive pressure transducers with integrated circuits, *Sensors and Actuators, 4* (1983) 403–411	pressure	2 × 4	6
Smith, Prisbe, Shott and Meindl, Integrated circuits for a capacitive pressure sensor, *Proc. IEEE Frontiers Eng. Comp. in Health Care*, 1984, pp. 440–443	pressure	2 × 6	22
Hanneborg and Ohlkers, A capacitive silicon pressure sensor with low TCO and high long-term stability, *Sensors and Actuators, A21-A23* (1990) 151–154	pressure	4 × 4	?
Smith, Bowman and Meindl, Analysis, design, and performance of capacitive pressure sensor IC, *IEEE Trans. Biomed. Eng., BME-33* (1986) 163–174	pressure	2 × 6	22
Chau and Wise, An ultra-miniature solid-state pressure sensor for a cardiovascular catheter, *Proc. 4th Int. Conf. Solid-State Sensors and Actuators (Transducers '87), Tokyo, Japan,* June 2–5, 1987, pp. 344–347	pressure	1 × 5	0.5
Miyoshi, Akiyama, Shintaku, Inami and Hijikigawa, A new fabrication process for capacitive pressure sensors, *Proc. 4th Int. Conf. Solid-State Sensors and Actuators (Transducers '87), Tokyo, Japan,* June 2–5, 1987, pp. 309–311	pressure	?	8
Shoji, Nisase, Esashi and Matsuo, Fabrication of an implantable capacitive type pressure sensor, *Proc. 4th Int. Conf. Solid-State Sensors and Actuators (Transducers '87), Tokyo, Japan,* June 2–5, 1987, pp. 305–308	pressure	2 × 3	10
Furuta, Esashi, Shoji and Matsumoto, Catheter-tip capacitive pressure sensor, *Tech. Digest, 8th Sensor Symp., Japan,* 1989, pp. 25–28	pressure	3.5 × 0.7	3.5
Puers, Peeters, Vanden Bossche and Sansen, A capacitive pressure sensor with low impedance output and active suppression of parasitic effects, *Sensors and Actuators, A21-23* (1990) 108–114	pressure	2 × 3.5	10
Bäcklund, Rosengren, Hök and Svedbergh, Passive silicon transensor intended for biomedical, remote pressure monitoring, *Sensors and Actuators, A21–A23* (1990) 58–61	pressure	3 × 3	25
Kandler, Eichholz, Manoli and Mokwa, CMOS compatible capacitive pressure sensor with read-out electronics, H. Reichl (ed.), *Microsystems,* Springer, New York, 1990, pp. 574–580	pressure	array 81 × 100 μm	2
Matsumoto, Shoji and Esashi, An integrated miniature capacitive pressure sensor, *Sensors and Actuators, A29* (1991) 185–193	pressure	2 × 1.7	
Puers, Peeters, VanDen Bossche, and Sansen, Harmonic response of silicon capacitive pressure sensor, *Sensors and Actuators, A25-27* (1991) 301–305	pressure	2 × 3.5	10
Puers, VanDen Bossche, Peeters and Sansen, An implantable pressure sensor for use in cardiology, *Sensors and Actuators, A21-23* (1990) 944–947	pressure	1.8 × 2.2	10
Artyomov, Kudryashov, Shelenshkevich and Shulga, Silicon capacitive pressure transducer with increased modulation depth, *Sensors and Actuators, A28* (1991) 223–230	pressure	4.5 × 4.5	7

Etch	Seal	Circuit	Application	Discussion
bulk	anodic	Schmitt oscillator, bipolar	cardio	early device
bulk, hydrazine	anodic	integrated, bipolar	general	ring vs. square membrane, drift aspects
bulk, KOH	anodic	separate	general	effect of sealing on TCO and TCS
bulk, hydrazine	anodic	integrated, CMOS	biomedical	ref. capacitor integrated with sensing capacitor
bulk	anodic	integrated, oscillator, bipolar 10 μm	cardio	temperature compensation
bulk	sputtered glass	separate chip	general	low TCO and drift
bulk	anodic	integrated, oscillator, bipolar 10 μm	cardio	comparison with piezoresistive devices
EDP, bulk	anodic	separate	cardio	miniaturized
surface etch	n.a.	no	general	Ni diaphragm, sacrificial layer, cheap large batch prod.
bulk, EDP	Si-Si fusion	CMOS	biomedical	direct fusion bonding
bulk, KOH	anodic	separate, CMOS	cardio	small assembly, compete backside etch
bulk, KOH	anodic	no	general	FEM analysis & linearization
KOH, bulk	fusion bonding	*LC* circuit only	eye-pressure	*LC* tuned by pressure-transponder system
sacrificial layer, polysilicon	n.a.	SC CMOS	general	preliminary results
bulk, KOH	anodic	integrated, CMOS	cardio	throughhole connection, backside etch
bulk, KOH	anodic	separate, CMOS	biomedical	parasitic capacitance rejection
bulk, KOH	anodic	separate	cardio	miniaturized biocompatible package
bulk	?	no	general	three electrode, linearization by reducing electrode

Authors	Type	Size (mm^2)	C_0 (pF)
Ji, Cho, Zhang, Najafi and Wise, An ultraminiature CMOS pressure sensor for a multiplexed cardiovascular catheter, *Proc. 6th Int. Conf. Solid-State Sensors and Actuators (Transducers '91), San Francisco, CA, USA*, June 24–28, 1991, pp. 1018–1020	pressure	1.4×0.4	0.3
Kudoh, Shoji and Esashi, An integrated miniature capacitive pressure sensor, *Sensors and Actuators, A29* (1991) 185–193	pressure	2.3×3.7	
Kung and Lee, An integrated air-gap capacitor process for sensor applications, *Proc. 6th Int. Conf. Solid-State Sensors and Actuators (Transducers '91), San Francisco, CA, USA*, June 24–28, 1991, pp. 1010–1013	pressure	0.4×0.5	
Nagata, Terabe, Fukaya, Sakurai, Tabata, Sugiyama and Esashi, Digital compensated capacitive pressure sensor using CMOS technology for low pressure measurements, *Proc. 6th Int. Conf. Solid-State Sensors and Actuators (Transducers '91), San Francisco, CA, USA,* June 24–28, 1991, pp. 308–311	pressure	5×5	
Rosengren, Söderkvist and Smith, Micromachined sensor structures with linear capacitive response, *Sensors and Actuators, A31* (1992) 200–205	pressure	2×2	3
Schnatz, Schöneberg, Brockherde, Kopystynski, Mehlhorn, Obermeirer and Benzel, Smart CMOS capacitive pressure transducer with on-chip calibration capability, *Sensors and Actuators, A34* (1992) 77–83	pressure	8.4×6.2	6
Suminto, Yeh, Spear and Ko, Silicon diaphragm capacitive sensor for pressure, flow, acceleration and altitude measurements, *Proc. 4th Int. Conf. Solid-State Sensors and Actuators (Transducers '87), Tokyo, Japan*, June 2–5, 1987, pp. 336–339	press., accel.	2×3	8
Puers and Vergote, A subminiature capacitive movement detector using a composite membrane suspension, *Sensors and Actuators, A31* (1992) 90–96	motion	1×1	4
Petersen, Shatel and Raley, Micromechanical accelerometer integrated with MOS detection circuitry, *IEEE Trans. Electron Devices, ED-29* (1982) 23–27	accel.	0.3×0.1	0.004
Rudolf, A micromechanical capacitive accelerometer with a two-point inertial mass suspension, *Sensors and Actuators, 4* (1983) 191–198	accel.	1.5×2.6	1
Rudolf, Jornod, and Bencze, Silicon microaccelerometer, *Proc. 4th Int. Conf. Solid-State Sensors and Actuators (Transducers '87), Tokyo, Japan*, June 2–5, 1987, pp. 395–398 Rudolf, Jornod, Bergqvist, and Leuthold, Precision accelerometers with µg resolution, *Sensors and Actuators, A21-23* (1990) 297–302	accel.	8×6	20
Olney, Acceleration measurement using variable capacitance, *Proc. Sensors Nüremberg '88, Germany*, 1988, pp. 149–160	accel.	2.8×3.6	10
Schlaak, Arndt, Steckenborn, Gevatter, Kiesewetter and Grethen, Micromechanical capacitive acceleration sensor with force compensation, H. Rechl (ed.), *Microsystems*, Springer, New York, 1990, pp. 617–622	accel.	7×7	16
Seidel, Riedel, Kolbeck, Mück, Kupke and Königer, Capacitive silicon accelerometer with highly symmetrical design, *Sensors and Actuators, A21-23* (1990) 312–315	accelerometer	35×3.5	10
Suzuki, Tuchitani, K. Sato, Ueno, Yokata, M. Sato and Esashi, Semiconductor capacitance-type accelerometer with PWM electrostatic servo technique, *Sensors and Actuators, A21-23* (1990) 316–319	accelerometer	3×4.5	9
Kloeck, Suzuki, Tuchitani, Miki, Matsumoto, K. Sato, Koide and Sugisawa, Motion investigation of electrostatic servo-accelerometers by means of transparent ITO fixed electrodes, *Proc. 6th Int. Conf. Solid-State Sensors and Actuators (Transducers '91), San Francisco, CA, USA*, June 24–28, 1991, pp. 108–111	accelerometer	3×4.5	9

Etch	Seal	Circuit	Application	Discussion
bulk	anodic	separate	cardio	subminiature assembly
bulk, KOH	anodic	integrated CMOS oscillator	general	advanced assembly, electrical feedthroughs
surface etch, KOH through	n.a.	integrated, NMOS	general	polydiaphragm, miniature size
bulk, TMAH	anodic	integrated, CMOS oscillator	general	frequency output, linearized response
bulk, KOH	fusion bond	no	test	linearization techniques by constructions
bulk	anodic	integrated, CMOS SC	general	includes band-gap reference for temp. correction
bulk, EDP	anodic	separate, CMOS	general	membrane with central mass for linearity
KOH, bulk	anodic	separate	animal monitoring	composite suspension membrane
surface etch, EDP	n.a.	one CMOS stage	general	silicon oxide beam, with Au deposited mass
KOH bulk	anodic	separate	general	plate suspended on torsion bars
KOH bulk	anodic	separate	spacecraft	force balancing, μg resolution
bulk	anodic	separate	space, flight control	commercial device
bulk, EDP	anodic	separate	position sensing	μg resolution, symmetric suspension
bulk	anodic	separate	general	good linearity
bulk	anodic	separate	general	servo action
bulk	anodic	separate	general	transparent electrodes on glass allow observation

Authors	Type	Size (mm^2)	C_0 (pF)
Payne and Dinswood, Surface micromachined accelerometer: a technology update, *SAE Int. Automotive Eng. Congr., Detroit, MI, USA,* 1991, pp. 127–135 Goodenough, Airbags boom when IC accelerometer sees 50g, *Electronic Design* (8 Aug. 1991), 127–135	accelero-meter	±5 × 5	?
Peeters, Vergote, Puers and Sansen, A highly symmetrical capacitive micro-accelerometer with single degree of freedom response, *Proc. 6th Int. Conf. Solid-State Sensors and Actuators (Transducers '91), San Francisco, CA, USA,* June 24–28, 1991, pp. 97–100	accelero-meter	3.6 × 3.6	12
Ura and Esashi, Differential capacitive accelerometer, *Tech. Dig. 10th Sensor Symp., Japan,* 1991, pp. 41–44	accelero-meter	5 × 6	15 × 2

Etch	Seal	Circuit	Application	Discussion
surface etch	n.a.	integrated CMOS	automotive	fully integrated device, no bulk micromachining
bulk, KOH	fusion bond, anodic	separate	general	highly symmetrical, tunable damping
bulk, hydrazine	anodic	separate	general	silicon-oxinitride suspension

Appendix 2

Dielectric Properties
of Various Materials

Excerpted, by permission of the publisher, from von Hippel, ed. *Dielectric Materials and Applications.* ©1954 by The MIT Press.

Solids, Inorganic — Crystals	°C	1×10^2 Hz ε_r	$\tan \delta \times 10^3$	1×10^3 ε_r	$\tan \delta \times 10^3$	1×10^4 ε_r	$\tan \delta \times 10^3$	1×10^5 ε_r	$\tan \delta \times 10^3$
Ice, from conductivity water	−12							4.8	800
Snow, freshly fallen	−20			3.33	492	1.82	342	1.24	140
Snow, hardpacked followed by light rain	−6							1.9	1530
Aluminum oxide, sapphire Field ⊥ optical axis	25	8.6	< 1	8.6	< 0.2	8.6	< 1	8.6	< 1
Field ∥ optical axis	25	10.55	< 1	10.55	<0.2	10.55	< 1	10.55	< 1
Ammonium dihydrogen phosphate Field ⊥ optical axis	25	56.4	40	56.0	4.6	55.9	0.46	55.9	< 0.5
Field ∥ optical axis	25	16.4	240	16.0	24	15.4	7	14.7	7
Lithium fluoride	25	9.00	1.5	9.00	< 0.3	9.00	< 0.2	9.00	< 0.2
	80	9.11	12.0	9.11	2.0	9.11	1.1	9.11	0.4
Magnesium oxide	25	9.65	< 0.3	9.65	< 0.3	9.65	< 0.3	9.65	< 0.3
Potassium bromide	25	4.90	0.7	4.90	0.7	4.90	0.8	4.90	0.45
	87	4.97	1.6	4.97	0.7	4.97	1.1	4.97	0.9
Potassium dihydrogen phosphate Field ⊥ optical axis	25	44.5	9.8	44.3	1.5	44.3	< 0.5	44.3	< 0.5
Field ∥ optical axis	25	21.4	17.0	20.7	2.4	20.5	< 2	20.3	< 0.5
Selenium, multicrystalline	25								
Sodium chloride, fresh crystals [Harshaw]	25	5.90	< 0.1	5.90	< 0.1	5.90	< 0.1	5.90	< 0.2
	85	6.35	17.0	6.11	24.0	6.00	7.0	5.98	0.6
Sulfur, sublimed (U.S.P.)	25	3.69	0.3	3.69	0.2	3.69	< 0.2	3.69	< 0.2
Thallium bromide	25	31.1	130	30.3	12.8	30.3	1.33	30.3	0.2
Thallium iodide	25	22.3	95	21.8	12	21.8	1.2	21.8	0.12
	193								
Titanium dioxide, rutile Field ⊥ optical axis	25	87.3	11	86.7	3.2	86.4	0.9	86	0.4
Field ∥ optical axis	25					200	350	170	60

Solids, Inorganic — Ceramics	°C	1×10^2 Hz ε_r	$\tan \delta \times 10^3$	1×10^3 ε_r	$\tan \delta \times 10^3$	1×10^4 ε_r	$\tan \delta \times 10^3$	1×10^5 ε_r	$\tan \delta \times 10^3$
AlSiMag A-35	23	6.10	15	5.96	10	5.89	7	5.86	5
	85	6.84	89	6.37	37	6.11	17.5	5.96	10.3
Ceramic F-66	25	6.22	1.45	6.22	0.9	6.22	0.5	6.22	0.2
Steatite Type 302	25	5.80	3.2	5.80	2.0	5.80	1.6	5.80	1.3
Steatite Type 400	25	5.54	16	5.54	10	5.54	7.2	5.54	6.0
Steatite Type 410	25	5.77	5.5	5.77	3.0	5.77	1.6	5.77	0.9
Steatite Type 452	25	8.15	6.5	8.15	2.8	8.15	1.7	8.15	1.2
Porcelains Ziroconium porcelain Zi-4	25	6.44	5.9	6.40	4.0	6.35	3.1	6.32	2.7

1×10^6		1×10^7		1×10^8		3×10^8		3×10^9		1×10^{10}		2.5×10^{10}	
ε_r	$\tan \delta \times 10^3$	ε_r	$\tan \delta \times 10^3$	ε_r	$\tan \delta \times 10^3$	ε_r	$\tan \delta \times 10^3$	ε_r	$\tan \delta \times 10^3$	ε_r	$\tan \delta \times 10^3$	ε_r	$\tan \delta \times 10^3$
4.15	120	3.7	18					3.20	0.9	3.17	0.7		
1.20	21.5	1.20	4			1.20	1.2	1.20	0.29	1.26	0.42		
1.55	290							1.5	0.9				
8.6	< 1	8.6	< 1	8.6	< 1	8.6	< 0.1					8.6	1.4
10.55	< 1	10.55	< 1	10.55	< 1	10.55	< 0.1						
55.9	< 0.5	55.9	< 0.5	55.9	< 0.5	55.9	< 1.0						
14.3	6	14.3	1			14.3	0.5			13.7	5		
9.00	< 0.2	9.00	< 0.2			9.00	0.07			9.00	0.18		
9.11	< 0.2	9.11	< .0.2							9.11	0.33		
9.65	< 0.3	9.65	< 0.3	9.65	< 0.3								
4.90	< 0.2	4.90	< 0.2			4.90	< 0.1			4.90	0.23		
4.97	0.5	4.97	0.3			4.97	0.24			4.97	0.35		
44.3	< 0.5	44.3	< 0.5	44.3	< 0.5								
20.2	< 0.5	20.2	< 0.5	20.2	< 0.5								
						11.0	250	10.4	154			7.5	110
5.90	< 0.2	5.90	< 0.2									5.90	< 0.5
5.98	< 0.2	5.98	< 0.2									5.97	< 0.39
3.69	< 0.2	3.69	< 0.2					3.62	0.04	3.58	0.015		
30.3	0.1	30.3	0.04										
21.8	0.05	21.8	0.05										
		37.3	82										
85.8	0.2	85.8	0.2										
170	8	160	1.6										

1×10^6		1×10^7		1×10^8		3×10^8		3×10^9		1×10^{10}		2.5×10^{10}	
ε_r	$\tan \delta \times 10^3$	ε_r	$\tan \delta \times 10^3$	ε_r	$\tan \delta \times 10^3$	ε_r	$\tan \delta \times 10^3$	ε_r	$\tan \delta \times 10^3$	ε_r	$\tan \delta \times 10^3$	ε_r	$\tan \delta \times 10^3$
5.84	3.8	5.80	3.5	5.75	3.7			5.6	4.1			5.36	5.8
5.86	7.7	5.80	5.0	5.75	5.0			5.50	4.7				
6.22	0.1	6.22	0.15	6.22	0.3			6.22	0.55			6.2	1.1
5.80	1.2	5.80	1.2	5.80	1.2			5.80	1.9	5.80	3.6		
5.54	5.0	5.54	4.5	5.54	3.9			5.5	3.9	5.5	5.3		
5.77	0.7	5.77	0.6	5.77	0.6			5.7	0.89	5.7	2.2		
8.15	1.0	8.15	1.0	8.15	1.0			8.15	2.0	8.15	3.10		
6.32	2.3	6.30	2.1	6.30	2.5	6.30	2.7	6.23	4.5	6.18	5.7		

Solids, Inorganic		1×10^2 Hz		1×10^3		1×10^4		1×10^5	
Glasses	°C	ε_r	tan δ $\times 10^3$	ε_r	tan δ $\times 10^3$	ε_r	tan δ $\times 10^3$	ε_r	tan δ $\times 10^3$
Porcelain, wet process	25	6.47	28	6.24	18	6.08	13	5.98	10.5
Porcelain, dry process	25	5.50	22	5.36	14	5.23	10.5	5.14	8.5
Coors AI-200	25	8.83	1.4	8.83	0.57	8.82	0.48	8.80	0.38
Porcelain #4462	25	8.99	2.2	8.95	0.91	8.95	0.60	8.95	0.30
Coors AB-2	25	8.22	2.0	8.18	1.34	8.17	1.14	8.17	1.05
AlSiMag 491	25								
Miscellaneous ceramics									
Beryllium oxide	25	4.61	17	4.47	8.4	4.41	7.4	4.34	7.2

Solids, Inorganic		1×10^2 Hz		1×10^3		1×10^4		1×10^5	
Glasses	°C	ε_r	tan δ $\times 10^3$	ε_r	tan δ $\times 10^3$	ε_r	tan δ $\times 10^3$	ε_r	tan δ $\times 10^3$
Corning 8871	25	8.45	1.8	8.45	1.3	8.45	0.9	8.45	0.7
Corning 9010	25	6.51	5.05	6.49	3.62	6.48	2.67	6.45	2.27
Corning Lab. No. 189CS	25	19.2	1.25	19.2	1.3	19.2	1.65	19.1	2.1
"E" glass	23	6.43	4.2	6.40	3.4	6.39	2.7	6.37	1.8
Foamglas	23	90.0	150	82.5	160	68.0	238	44.0	320
Fused silica 915	25	3.78	0.66	3.78	0.26	3.78	0.11	3.78	0.04
Fused quartz	25	3.78	0.85	3.78	0.75	3.78	0.6	3.78	0.4
Soda-silica glasses									
9% Na_2O, 91% SiO_2	25	6.4	250	6.2	82	5.7	40		
20% Na_2O, 80% SiO_2	25	10.8	400	8.3	150	7.3	67	6.8	36
30% Na_2O, 70% SiO_2	25	18	1100	12	390	10.4	130		
Alkali-silica glasses									
12.8% Li_2O, 87.2% SiO_2	25	9.94	970	6.54	360	5.45	100	5.1	31
12.8% Na_2O, 87.2% SiO_2	25	8.09	305	6.61	137	6.00	45	5.8	24
12.8% K_2O, 87.2% SiO_2	25	7.53	50.2	6.49	36	6.25	20	6.17	12.1
12.8% Rb_2O, 87.2% SiO_2	25	5.39	9.8	5.32	8.9	5.23	5.8	5.22	4.6
6.4% Li_2O, 6.4% Na_2O, 87% SiO_2	25	5.15	14.5	5.08	8.7	5.05	4.7	5.05	2.8
3.3% Li_2O, 6.6% K_2O, 71% SiO_2	25	5.23	5.3	5.19	4.7	5.17	3.7	5.15	2.8
6.4% Na_2, 6.4% K_2O, 87.2% SiO_2	25	5.68	10.2	5.62	7.5	5.58	4.2	5.56	3.1

Solids, Inorganic		1×10^2 Hz		1×10^3		1×10^4		1×10^5	
Miscellaneous	°C	ε_r	tan δ $\times 10^3$	ε_r	tan δ $\times 10^3$	ε_r	tan δ $\times 10^3$	ε_r	tan δ $\times 10^3$
Ruby mica (muscovite)	26	5.4	2.5	5.4	0.6	5.4	0.35	5.4	0.3
Canadian mica									
(field ⊥ sheet)	25	6.90	1.5	6.90	0.2	6.90	0.1		
(field ∥ sheet)	25	11.5	230	8.7	98	7.3	40		
Marble S-3030 (after drying)	25	15.6	200	12.8	110	11.4	63	10.6	39
Selenium, amorphous	25	6.00	1.8	6.00	0.4	6.00	<0.3	6.00	<0.5
Sandy soil, dry	25	3.42	.0196	2.91	0.008	2.75	.0034	2.65	0.002
2–18% moisture	25	3.23	0.064	2.72	0.013	2.50	.0056	2.50	0.003
3.88% moisture	25				150			5.0	1.9
16.8% moisture	25				342.5		36.7		

1×10^6		1×10^7		1×10^8		3×10^8		3×10^9		1×10^{10}		2.5×10^{10}	
ε_r	$\tan \delta \times 10^3$	ε_r	$\tan \delta \times 10^3$	ε_r	$\tan \delta \times 10^3$	ε_r	$\tan \delta \times 10^3$	ε_r	$\tan \delta \times 10^3$	ε_r	$\tan \delta \times 10^3$	ε_r	$\tan \delta \times 10^3$
5.87	9.0	5.82	11.5	5.80	13.5	5.75	14			5.51	15.5		
5.08	7.5	5.04	7.0	5.04	7.8	5.02	9.8			4.74	15.6		
8.80	0.33	8.80	0.32	8.80	0.30			8.79	1.0	8.79	1.8		
8.95	0.20	8.95	0.20	8.95	0.40			8.90	1.1	8.80	1.4		
8.16	0.9	8.16	0.75	8.16	0.9			8.14	1.6	8.08	2.7		
8.74	2.2							8.60	1.7	8.50	2.3		
4.28	3.8	4.24	1.9	4.23	1.25					4.20	0.5		

1×10^6		1×10^7		1×10^8		3×10^8		3×10^9		1×10^{10}		2.5×10^{10}	
ε_r	$\tan \delta \times 10^3$	ε_r	$\tan \delta \times 10^3$	ε_r	$\tan \delta \times 10^3$	ε_r	$\tan \delta \times 10^3$	ε_r	$\tan \delta \times 10^3$	ε_r	$\tan \delta \times 10^3$	ε_r	$\tan \delta \times 10^3$
8.45	0.6	8.43	0.7			8.40	1.4	8.34	2.6	8.05	4.9	7.82	7.0
6.44	2.15	6.43	2.26	6.42	3.0	6.40	4.1			6.27	9.1		
19.0	2.7	19.0	3.7	19.0	5.7			17.8	12.4				
6.32	1.5	6.25	1.7	6.22	2.3					6.11	6.0		
17.5	318	9.0	196							5.49	45.5		
3.78	0.01	3.78	0.01	3.78	0.03	3.78	0.05			3.78	0.17		
3.78	0.2	3.78	0.1	3.78	0.1			3.78	0.06	3.78	0.1	3.78	0.25
5.4	13					5.1	10			5.05	13	4.9	16
6.6	22	6.3	18			5.9	14			5.6	20	6.1	28
8.5	40					7.5	19			7.2	24	7.0	35
4.95	17.4	4.92	12.4			4.9	7.9			4.80	10.2		
5.66	15.9	5.57	12.6			5.4	11.8			5.33	18.2		
6.09	9	6.02	8			5.8	9.9			5.8	22		
5.21	4.1	5.20	3.8			5.15	5.9			5.05	12		
5.04	1.9	5.03	1.7			5.00	2.6			4.95	5.2		
5.14	2.4	5.10	2.4			5.07	4.0			5.04	8.3		
5.56	2.5	5.54	2.3			5.51	4.0			5.50	11.5		

1×10^6		1×10^7		1×10^8		3×10^8		3×10^9		1×10^{10}		2.5×10^{10}	
ε_r	$\tan \delta \times 10^3$	ε_r	$\tan \delta \times 10^3$	ε_r	$\tan \delta \times 10^3$	ε_r	$\tan \delta \times 10^3$	ε_r	$\tan \delta \times 10^3$	ε_r	$\tan \delta \times 10^3$	ε_r	$\tan \delta \times 10^3$
5.4	0.3	5.4	0.3	5.4	0.2			5.4	0.3				
10.0	36	9.5	37	9.1	29	8.8	25			8.6	12		
6.00	<3	6.00	<0.2	6.00	<0.2	6.00	<0.5	6.00	0.18	6.00	0.67	6.00	1.3
2.59	.0017	2.55	.0016			2.55	.0010	2.55	.00062	2.53	.00036		
2.50	.0025	2.50	.0025			2.50	.0026	2.50	0.003	2.50	0.0065		
4.70	0.175	4.50	0.03			4.50	0.003	4.40	.0046	3.60	0.012		
20	0.4	20	0.035			20	0.003	20	0.013	13	.029		

Solids, Inorganic Miscellaneous	°C	1×10^2 Hz		1×10^3		1×10^4		1×10^5	
		ε_r	tan δ $\times 10^3$	ε_r	tan δ $\times 10^3$	ε_r	tan δ $\times 10^3$	ε_r	tan δ $\times 10^3$
Loamy soil, dry	25	3.06	0.007	2.83	0.005	2.69	.0035	2.60	0.003
2.2% moisture	25				0.21	18	0.16		
13.77% moisture	25				849		97		
Clay soil, dry	25	4.73	0.012	3.94	0.012	3.27	0.012	2.79	0.01
20.09% moisture	25				780		100		

Solids, Organic Plastics	°C	1×10^2 Hz		1×10^3		1×10^4		1×10^5	
		ε_r	tan δ $\times 10^3$	ε_r	tan δ $\times 10^3$	ε_r	tan δ $\times 10^3$	ε_r	tan δ $\times 10^3$
Nylon 610	25	3.60	15.5	3.50	18.6	3.35	20.8	3.24	22.1
	84	13.5	235	11.2	140	9.0	158	6.3	203
Nylon 610, 90% humidity	25	4.5	65	4.2	64	4.0	60	3.7	50
Cellulose derivatives, acetates									
LL-l	25	3.82	9.5	3.77	15	3.67	20	3.53	23
	85	3.98	3.4	3.96	7.5	3.90	13.5	3.77	21
Silicon resins									
Dilecto (silicone glass laminate) 50%	25	3.56	1.35	3.56	1.29	3.56	1.29	3.55	1.11
DC2103, 50% glass, field ⊥ sheet	90	3.41	2.64	3.39	1.62	3.38	1.49	3.38	1.61
	200	3.18	23.6	3.18	4.3	3.18	2.3	3.17	19.5
Field ‖ sheet	25	6.55	125	5.72	91	4.94	108	3.96	51.6
	215	10.93	200	8.6	208	6.65	180	4.92	141
Measured after temperature run	25	5.71	62.4	5.16	88	4.54	94	3.98	64
Polyvinyl resins									
Polyethylene DE-3401	25	2.26	<0.2	2.26	<0.2	2.26	<0.2	2.26	<0.2
Polyvinyl chlorides									
Polyvinyl chloride W-176	25	6.21	73	5.52	94	4.70	107	3.96	96
Plasticell	25	1.04	2.1	1.04	1.1	1.04	<1.5	1.04	<1.5
Polyvinylidene and vinyl chloride									
Saran B-115	23	4.88	45	4.65	63	4.17	88.5	3.60	8.45
Polychlorotriflouroethylene									
Kel-F	26	2.72	21	2.63	27	2.53	23	2.46	13.5
Polytetraflouroethylene									
Dilecto (Teflon Laminate GB-112T, 67%	25	2.76	0.89	2.74	0.61	2.74	0.6	2.74	0.6
Tef., 33% glass), field ⊥ laminate	250	2.48	8	2.46	3.6	2.46	1.8	2.46	1.4
After heating	25	2.70	0.47	2.68	0.39	2.69	0.47	2.69	0.48
Polyacrylates									
Polyethyl methacrylate	22	2.90	42	2.75	29.4	2.65	18.5	2.60	11.8
	80	3.87	81	3.36	106	2.86	96	2.70	71
Polyisobutyl methacrylate	25	2.70	11.1	2.68	7.0	2.63	5.0	2.55	3.7
	80	2.9	83	2.7	60	2.5	36	2.5	21
Polystyrene									
Polystyrene (molded sheet stock)	25	2.56	<0.05	2.56	<0.05	2.56	<0.05	2.56	.05
	80	2.54	0.9	2.54	0.2	2.54	<0.1	2.54	<0.2

1×10^6		1×10^7		1×10^8		3×10^8		3×10^9		1×10^{10}		2.5×10^{10}	
ε_r	$\tan \delta \times 10^3$	ε_r	$\tan \delta \times 10^3$	ε_r	$\tan \delta \times 10^3$	ε_r	$\tan \delta \times 10^3$	ε_r	$\tan \delta \times 10^3$	ε_r	$\tan \delta \times 10^3$	ε_r	$\tan \delta \times 10^3$
2.53	.0018	2.48	.0014			2.47	.00065	2.44	.00011	2.44	.00014		
6.9	0.065	4	0.045			3.5	0.006	3.5	0.004	3.50	0.003		
		14.5	0.13			20	0.016	20	0.012	13.8	0.018		
2.57	.0065	2.44	0.004			2.38	0.002	2.27	.0015	2.16	0.0013		
		21.6	0.17			20	0.052	11.3	0.025				

1×10^6		1×10^7		1×10^8		3×10^8		3×10^9		1×10^{10}		2.5×10^{10}	
ε_r	$\tan \delta \times 10^3$	ε_r	$\tan \delta \times 10^3$	ε_r	$\tan \delta \times 10^3$	ε_r	$\tan \delta \times 10^3$	ε_r	$\tan \delta \times 10^3$	ε_r	$\tan \delta \times 10^3$	ε_r	$\tan \delta \times 10^3$
3.14	21.8	3.05	20.5	3.0	20			2.84	11.7			2.73	10.5
4.4	172	3.7	115	3.4	67			2.94	35.6				
3.2	38	3.05	28	3.0	22			2.85	12.5				
3.42	23	3.30	21	3.29	21	3.28	22	3.24	29	3.24	40	3.1	31
3.58	27	3.44	28					3.30	29				
3.54	0.85	3.54	1.14	3.54	1.86								
3.38	1.1	3.38	0.98										
3.15	2.1	3.10	1.54										
3.84	12.8	3.82	3.7	3.80	2.5	3.78	3.9	3.76	5.2	3.70	8.7		
										3.78	8		
2.26	<0.2	2.26	<0.2					2.26	0.31	2.26	0.36		
3.53	72	3.28	52	3.00	50								
1.04	1.0			1.04	1.0			1.04	5.5	1.04	5.0		
3.18	57	2.97	31	2.82	18			2.71	7.2	2.70	5.1		
2.43	8.2	2.35	6.0			2.30	3.0	2.30	2.8	2.29	3.9	2.29	5.5
2.73	0.58	2.73	0.62	2.73	1.18								
		2.44	0.25										
2.55	9.0	2.53	7.5					2.51	7.5	2.50	9.7	2.5	8.3
2.61	40	2.57	26	2.55	14			2.49	9.1	2.48	13.5		
2.45	3.5	2.45	4.6	2.42	5.2	2.40	4.7	2.39	3.1	2.38	3.9	2.37	5.2
2.44	10	2.42	8.0	2.42	7.0	2.40	6.5	2.39	5.4	2.38	5.9		
2.56	0.07	2.56	<0.2	2.55	<0.1	2.55	0.35	2.55	0.33	2.54	0.43	2.54	1.2
2.54	<0.2	2.54	<0.2	2.54	<0.3	2.54	0.27	2.54	0.45	2.53	0.53		

Solids, Organic Plastics	°C	1 × 10² Hz		1 × 10³		1 × 10⁴		1 × 10⁵	
		ε_r	tan δ × 10³	ε_r	tan δ × 10³	ε_r	tan δ × 10³	ε_r	tan δ × 10³
Formica LE (after 5 yr. storage)	25	5.92	.0112	5.32	.0057	5.01	.0038	4.86	.004
	100			16.0	.042	8.16	.029	6.22	.0151
Micarta 299	200	12.8	.059	6.7	.036	5.0	.0158	4.75	.0046
Mycalex 400, dry	25	7.45	.00029	7.45	.00019	7.42	.00016	7.40	.00014
after 48 h. in H₂O	25	7.45	.0012	7.45	.00097	7.42	.00068		
Bakelite BM 262, dry	25	4.85	.00095	4.80	.00082	4.74	.00075	4.72	.00060
after 20 days, 90% rel. hum	25	3.87	.00184	3.78	.0015	3.70	.0012		
after 7 mos., 90% rel. hum	25	6.72	.0147	5.78	.009	5.15	.0057		
after 19 mos., 90% rel. hum.	25	9.10	.0244	6.75	.0169	5.51	.0111		
Formica FF-55, dry	25								
after 20 h. in H₂O	25								
GMG melamine, dry	25	8.2	.019	7.0	.0069	6.7	.0019	6.6	.001
after 6 or 8 mos., 90% rel. hum.	25	42.5	.075	16.8	.054	10.4	.027	7.65	.010
Polythene, dry	25								
after 10 days, 90% rel. hum.	25								

Solids, Organic Miscellaneous	°C	1 × 10² Hz		1 × 10³		1 × 10⁴		1 × 10⁵	
		ε_r	tan δ × 10³	ε_r	tan δ × 10³	ε_r	tan δ × 10³	ε_r	tan δ × 10³
Paper, Royalgrey	25	3.30	5.8	3.29	7.7	3.22	11.7	3.10	20
	82	3.57	17	3.52	7.4	3.49	6.1	3.40	8.5
Leather, sole, dried	25	4.1	45	3.9	35	3.6	30	3.4	28
Leather, sole, about 15% moisture	25	38	1400	14.0	700	9.3	370	6.9	220
Soap, Ivory	25								
Steak (bottom round)	25					24,400	40,500		
Suet	25			750	3000	210	2500		

Liquids, Inorganic Miscellaneous	°C	1 × 10² Hz		1 × 10³		1 × 10⁴		1 × 10⁵	
		ε_r	tan δ × 10³	ε_r	tan δ × 10³	ε_r	tan δ × 10³	ε_r	tan δ × 10³
Water, conductivity	1.5							87.0	190
	5								
	15								
	25							78.2	400
	35								
	45								
	55								
	65								
	75								
	85							58	1240
	95								
Aqueous sodium chloride									
0.1 molal solution	25							78.2	2.4·10⁶
0.3 molal solution	25							78.2	6.3·10⁶

1×10^6		1×10^7		1×10^8		3×10^8		3×10^9		1×10^{10}		2.5×10^{10}	
ε_r	$\tan\delta \times 10^3$	ε_r	$\tan\delta \times 10^3$	ε_r	$\tan\delta \times 10^3$	ε_r	$\tan\delta \times 10^3$	ε_r	$\tan\delta \times 10^3$	ε_r	$\tan\delta \times 10^3$	ε_r	$\tan\delta \times 10^3$
4.85	.0042							3.63	.0037				
5.84	.008							4.10	.0067				
4.6	.0015							4.4	.0016				
7.39	.00013	7.38	.00013							7.12	.00033		
7.39	.00029	7.38	.00021							7.18	.0010		
4.67	.00055												
4.18	.0047												
						5.55	.0027						
						6.5	.0064						
6.4	.0011												
6.57	.0030												
								2.26	$67 \cdot 10^{-6}$				
								2.26	$85 \cdot 10^{-6}$				

1×10^6		1×10^7		1×10^8		3×10^8		3×10^9		1×10^{10}		2.5×10^{10}	
ε_r	$\tan\delta \times 10^3$	ε_r	$\tan\delta \times 10^3$	ε_r	$\tan\delta \times 10^3$	ε_r	$\tan\delta \times 10^3$	ε_r	$\tan\delta \times 10^3$	ε_r	$\tan\delta \times 10^3$	ε_r	$\tan\delta \times 10^3$
2.99	38	2.86	57	2.77	66	2.75	66	2.70	56	2.62	40.3		
3.31	23	3.14	44	3.08	63	3.00	72	2.94	80	2.84	82.7		
3.2	28	3.1	30	3.1	38								
5.6	140	4.9	100	4.5	100								
								2.96	176.5				
197	61,000	50	26,000			50	780	40	300	30	370	15	400
14	1700	4.5	930	2.6	150	2.5	120	2.5	70	2.5	50	2.4	50

1×10^6		1×10^7		1×10^8		3×10^8		3×10^9		1×10^{10}		2.5×10^{10}	
ε_r	$\tan\delta \times 10^3$	ε_r	$\tan\delta \times 10^3$	ε_r	$\tan\delta \times 10^3$	ε_r	$\tan\delta \times 10^3$	ε_r	$\tan\delta \times 10^3$	ε_r	$\tan\delta \times 10^3$	ε_r	$\tan\delta \times 10^3$
87.0	19	87	2.0	87	7.0	86.5	32	80.5	310	38	1030	15	425
85.5	22					85.2	27.3	80.2	275	41	950	17.5	395
81.7	31					81.0	21	78.8	205	49	700	25	330
78.2	40	78.2	4.6	78	5.0	77.5	16	76.7	157	55	540	34	265
74.8	48.5					74.0	12.5	74.0	127	58	440	41	215
71.5	59					71.0	10.5	70.7	106	59	400	46	275
68.2	72					68	9.2	67.5	89	60	360	49	245
64.8	86.5					64.5	8.4	64.0	76.5	59	320	50.5	125
61.5	103					61	7.7	60.5	66	57	280	51.5	105
58	124	58	12.5	58	3.0	57	7.3	56.5	54.7	54	260		
55	143					52	7.0	52	47				
						76	780	75.5	240	54	560		
						71	2400	69.3	435	52	605		

Liquids, Inorganic Miscellaneous	°C	1×10^2 Hz ε_r	tan δ $\times 10^3$	1×10^3 ε_r	tan δ $\times 10^3$	1×10^4 ε_r	tan δ $\times 10^3$	1×10^5 ε_r	tan δ $\times 10^3$
0.5 molal solution	25							78.2	$9.9 \cdot 10^6$
0.7 molal solution	25							78.2	$13 \cdot 10^6$

Liquids, Organic	°C	1×10^2 ε_r	tan δ $\times 10^3$	1×10^3 ε_r	tan δ $\times 10^3$	1×10^4 ε_r	tan δ $\times 10^3$	1×10^5 ε_r	tan δ $\times 10^3$
Aliphatic									
Heptane	25	1.971	<0.3	1.971	<0.04	1.971	<0.04		
Methyl alcohol	25								
Ethyl alcohol	25								
Ethylene glycol	25					42	3000	41	300
Carbon tetrachloride	25	2.17	6.0	2.17	0.8	2.17	0.04	2.17	<0.04
Tetrachloroethylene	25	2.28	1.5	2.28	0.2	2.28	0.07	2.28	0.1
Dichloropentanes #40	25			334	520,000	17.1	106,000	8.65	13,500
Aromatic									
HB-40 oil	25	2.59	0.13	2.59	<0.04	2.59	<0.04	2.59	<0.3
Pyranol 1467	25	4.42	3.6	4.40	0.3	4.40	<0.4	4.40	0.36
Petroleum oils									
Aviation gasoline, 100 octane	25					1.94	0.1		
Aviation gasoline, 91 octane	25					1.95	0.4		
Jet fuel JP-1	25					2.12	<0.1		
Jet fuel JP-3	25					2.08	<0.1		
Vaseline	25	2.16	0.3	2.16	0.2	2.16	<0.2	2.16	<0.1
	80	2.10	1.6	2.10	0.36	2.10	.09	2.10	<0.1
Cable oil 5314	25	2.25	0.3	2.25	<0.04	2.25	<0.04	2.25	<0.1
	80	2.18	3.8	2.18	0.4	2.18	0.05		
Silicones									
DC500, 0.65 cs. at 25°C	−15	2.20	<0.5	2.20	<0.3	2.20	<0.3	2.20	<0.5
	22	2.20	0.1	2.20	<0.04	2.20	<0.04	2.20	<0.3
DC500, 10 cs. at 25°C	23	2.66	1.2	2.66	0.15	2.66	<0.04	2.66	<0.3
Ignition Sealing Compound #4 (organosiloxane polymer, Dow)	25	2.75	1.5	2.75	0.6	2.75	0.5	2.74	0.4
SF96-40 (General Electric)	25	2.71	<0.03	2.71	<0.003	2.71	<0.003	2.71	<0.03

1×10^6		1×10^7		1×10^8		3×10^8		3×10^9		1×10^{10}		2.5×10^{10}	
ε_r	$\tan \delta \times 10^3$	ε_r	$\tan \delta \times 10^3$	ε_r	$\tan \delta \times 10^3$	ε_r	$\tan \delta \times 10^3$	ε_r	$\tan \delta \times 10^3$	ε_r	$\tan \delta \times 10^3$	ε_r	$\tan \delta \times 10^3$
						69	3900	67.0	625	51	630		
										50	660		

1×10^6		1×10^7		1×10^8		3×10^8		3×10^9		1×10^{10}		2.5×10^{10}	
ε_r	$\tan \delta \times 10^3$	ε_r	$\tan \delta \times 10^3$	ε_r	$\tan \delta \times 10^3$	ε_r	$\tan \delta \times 10^3$	ε_r	$\tan \delta \times 10^3$	ε_r	$\tan \delta \times 10^3$	ε_r	$\tan \delta \times 10^3$
						1.97	<0.25	1.97	.1	1.97	1.6		
31	200	31.0	26	31.0	38	30.9	80	23.9	640	8.9	810		
24.5	90	24.1	33	23.7	62	22.3	270	6.5	250	1.7	68		
41	30	41	8	41	45	39	160	12	1000	7	780		
2.17	<0.04	2.17	<0.2	2.17	<0.2	2.17	<0.1	2.17	0.4	2.17	1.6		
2.28	0.2	2.28	0.2					2.28	1.0				
		7.76	270			7.57	84	6.81	198				
2.58	1.3	2.57	7.6	2.54	16	2.48	9.3	2.40	3	2.34	1.7		
4.40	2.5	4.40	26	4.08	130	3.19	150	2.84	120	2.62	74		
						1.94	.08	1.92	1.4				
						1.95	.04	1.94	1.15				
						2.12	1.2	2.09	6.8				
						2.08	0.7	2.04	5.5				
2.16	<0.1	2.16	<0.3	2.16	<0.4			2.16	0.66	2.16	1.0		
2.10	<0.1							2.10	0.92	2.10	2.2		
						2.24	3.9	2.22	1.8	2.22	2.2		
								2.18	4.7				
2.20	<0.3	2.20	<0.2					2.20	1.86				
2.20	<0.3	2.20	<0.2			2.20	0.14	2.20	1.45	2.19	3.0	2.13	6.0
2.66	<0.3	2.66	0.3					2.65	6.8	2.63	27	2.48	41
2.75	0.4	2.75	0.6	2.74	1.5	2.72	2.8	2.65	9.2	2.49	27		
2.71	<0.1	2.71	<0.1			2.71	1.1	2.70	9.5	2.67	18.6		

Index